THE JESUS MICROBIOME
An Instagram from the First Century

Stephen J. Mattingly Ph.D.
Roy Abraham Varghese

Foreword by N.V. Perricone M.D.

Institute for MetaScientific Research,
Dallas, Texas
info@JesusMicrobiome.com

Institute for MetaScientific Research,
Dallas, Texas
info@JesusMicrobiome.com

© Copyright 2021
by Stephen J. Mattingly Ph.D. and Roy Abraham Varghese

Foreword copyright © 2021 by Nicholas V. Perricone

All Rights Reserved.
No part of this book may be reproduced, stored in a
retrieval system, or transmitted by any means,
electronic, mechanical, photocopying, recording,
or otherwise, without written permission
from the author.

ISBN Print: 978-1-7364447-0-2
ISBN eBook: 978-1-7364447-1-9

*T**he Jesus Microbiome* is a breakthrough book unveiling radically new but scientifically demonstrable, universally replicable and factually definitive findings on the most mysterious and controversial piece of cloth on earth: the Shroud of Turin. The implications of these never-before publicized findings are historic in nature for, as this book shows:

- The image on the Shroud of Turin was literally created from the flesh and blood of Christ and the underlying processes that led to this particular form of image creation can be replicated today.

- *Microbiome = the body of microorganisms that inhabit a specific environment.* Although well-intentioned, the Carbon 14 dating study performed on the Shroud over three decades ago (1988) was unavoidably flawed because the investigators were unaware of what is now known about microbiomes and failed to demicrobialize the Shroud. No decontamination = no valid dating.

- Since the Shroud incarnates pivotal events in the New Testament narratives in real-time it can rightly be described as the fifth Gospel – an Instagram from the first century sent by Jesus himself!

The Jesus Microbiome is not just another book about the Shroud. It is a journey into the first century through the 21st where the ancient world comes to life again through the medium of the microbiome.

Stephen J. Mattingly Ph.D. was Professor of Microbiology and Immunology at the University of Texas Health Science Center at San Antonio for 33 years and past President of the Texas Branch of the

American Society for Microbiology. He was first invited to participate in research on the Shroud of Turin in the 1990s and published several scientific papers and a book on his findings. Participants in his 1994 scientific conference on the Shroud included Harry Gove, co-inventor of the AMS Radiocarbon dating technology used in the 1988 Carbon 14 dating study, who had helped arrange the dating study. After his ongoing discussions with Mattingly, Gove later said in a BBC interview that microbial contamination "of the linen fibrils could not have been removed even by the most stringent pretreatment cleaning process and would, definitely, skew the real age of the linen."

Roy Abraham Varghese is the author and/or editor of sixteen books on the interface of science, philosophy, and religion. His *Cosmos, Bios, Theos*, included contributions from 24 Nobel Prize-winning scientists. *Time* magazine called *Cosmos* "the year's most intriguing book about God." *Cosmic Beginnings and Human Ends*, a subsequent work, won a Templeton Book Prize for "*Outstanding Books in Science and Natural Theology.*" Varghese's *The Wonder of the World* was endorsed by leading thinkers include two Nobelists and was the subject of an Associated Press story. He co-authored *There is a God— How the World's Most Notorious Atheist Changed His Mind* with Antony Flew. His most recent work, *The Missing Link*, a study of consciousness, thought and the human self, includes contributions from three Nobel Prize winners and scientists from Oxford, Cambridge, Harvard, and Yale.

Author of the Foreword – Nicholas V. Perricone M.D. Dr. Perricone, the "doctor to the stars," is the author of eight books on anti-aging, wellness, healthy skin, including three #1 New York Times bestsellers; a board-certified dermatologist named by *Vogue* as one of the four top dermatologists in the US; an award-winning inventor who holds more than 160 patents; a research scientist; and an internationally renowned expert on inflammation and systemic disease. He is the focus of a series of award-winning PBS specials and a popular guest on *Oprah, Good Morning America*, the *Today* show, and *20/20*, among others.

Acknowledgements

The authors thank the individuals listed below for their kind permission to quote extensively from their published works and/or the works to which they hold the rights. The works cited constitute an enduring contribution to the study of the Shroud of Turin.

Dr. Thomas de Wesselow
Dr. William Edward
Dr. Mark Guscin
Professor Daniel C. Scavone
Professor William Meacham
Mrs. Catherine Zugibe, wife of the late Dr. Frederick Zugibe
Mr. Garry Bucklin, son of the late Dr. Robert Bucklin

"They have closed their eyes, lest they see with their eyes"
Matthew 13:15

"Who are you going to believe – me or your own eyes?"
Comedian Chico Marx

Contents

A Microbiologist at Calvary xiii

Foreword xv

Prologue – *The Fifth Gospel* xix

Introduction – *From Relativity and Quantum Theory
    to the Shroud of Turin* xxi

Overview – *The Shroud* xxiv

Section I Shroud 3.0 1

    1.1 The Once and Future Shroud 3

    1.2 Shroud 2.0 Hard Facts 13

    1.3 Microbiome 101 21

    1.4 The Jesus Microbiome Project 25

    1.5 The Embedded Microbiome that Created the
         First Photograph in History 33

    1.6 Paradigm Shift a.k.a. Game-Changer 41

    1.7 It Seemed Like a Good Idea at the Time – Searching for
         the Carbon "Fingerprint" 45

    1.8 See for Yourself – Launching a Microbiome Here
         and Now 59

    1.9 Do it Yourself – Getting Your Microbiome to Create
         a Shroud-type Image 63

    1.10 Letting Sleeping Bugs Lie – Demonstrating the 1988
         Tests' Failure to Decontaminate 77

    1.11 FAQs – Critiques of the Microbial Paradigm and
         Responses 85

Section II Shroud in Toto **95**

    2.1 Based on a True Story 101

    2.2 Ecce Homo 115

    2.3 Doctors' Diagnoses 127

    2.4 Bloodstains 151

    2.5 "The stones will cry out" – Pollen, Dust, Cloth 161

    2.6 Sudarium and Shroud – a Tale of Two Images 179

    2.7 The Shroud – a History 189

- A Star is Born 189
- Three's a Shroud 196
- Shooting Stars 198
- Big Bang 207
- The Artifact Formerly Known as the Image of Edessa 212
- The Constantinople Constellation 219
- "I don't believe the Shroud is a medieval work of art because I don't believe in miracles" 231
- Starstruck 239

    2.8 Authenticity and Antiquity 241

Section III Who Was the Man of the Shroud? **247**

    3.1 Life, Death and Resurrection 251

    3.2 Connecting the Dots 265

Section IV Shroud in a Shroud **273**

    4.1 Guilty Until Proven Guilty 277

    4.2 And then a Miracle Happened! Unnatural Shroud Origin Theories 279

    4.3 Mistakes were Made … 287

Section V Shroud Unshrouded **293**
    5.1   Seeing is Believing 295
    5.2   Here and Now – Suffering and the Shroud 305

End Notes.. 307
Appendix 1: Photographs of the Microbiome Experiments............. 319
Appendix 2: Historical Documents supporting the Presence
of the Shroud in Constantinople
(from Professor Daniel C. Scavone) 323

The Shroud of Turin*

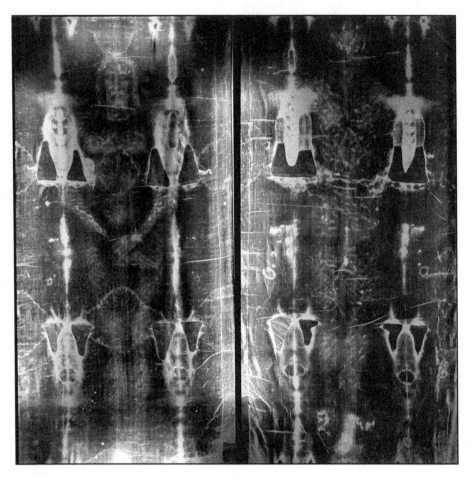

A Microbiologist at Calvary

Once Jesus was nailed to the cross, his physical appearance began to change over time. The blood from the head wounds and the numerous abrasions from the scourging merged with the white frothy bacterial polysaccharides produced by skin bacteria most especially *Staphylococcus epidermidis*. At twelve noon, there were three hours remaining before his death or six doublings of the number of bacteria on his body. So every thirty minutes the bacteria doubled in number. Thousands of bacteria became multiple millions per square centimeter. The bacterial numbers literally exploded with time. The bacteria with their sticky polysaccharides oozed into his eyes and dripped from his nose and mouth. The sticky polysaccharides attracted flies and insects as well as dust particles and any debris in the air. At the time of his death, he was a massive bacterial culture, a microbiome, unlike any associated with human life.

Foreword
"The Jesus Microbiome"
N.V. Perricone, M.D.

The Shroud of Turin. The very phrase conjures up images of a numinous talisman piously preserved in the alleyways and ramparts of mystical Italy.

The Shroud, so its champions say, is "in" Turin but not "of" it. For in reality it is the burial cloth of Jesus the Nazarene.

But since the 1988 radiocarbon dating of the Shroud, mainstream opinion has taken it for granted that this revered relic is a creation of the Middle Ages.

Hence the current impasse. On the one side are the few who, on faith, profess belief in the sacred origin of the Shroud. On the other are the multitudes who profess belief in the method used to date the Shroud and consequently believe it to be of medieval origin. And never the twain shall meet.

Three decades after the dating study, however, much has happened in our apprehension of the organic world that compels us to revisit the whole question. The present book goes beyond the science vs. religion and faith vs. reason debates to study the origin and nature of the Shroud of Turin entirely in terms of science.

Most relevant of all, in this context, is the discovery of the microbiome: the microbiological universe hidden in plain sight within and without the human body. Although the role of microorganisms had been known in earlier decades, the idea of the *human* microbiome (the microorganisms living "off" us and on which we are dependent) and its subsequent mapping came together only in the 21st century. The human microbiome is made of various other microbiomes – the gut microbiome, the skin microbiome and others.

Truth be told, the last two decades of discovery relating to the microbiology of the human have been as revolutionary as the era of the unveiling of the quantum realm in physics. We have uncovered not just the microbiome but the epigenome, the changes in gene expression caused by mechanisms other than changes in the underlying DNA sequence.

My own work has centered on inflammation at the cellular level which has changed the scientific perception of human health and aging.

For the first time in history we are learning to differentiate between chronological years and biological years. We are unlocking the secrets to keeping our brains and our endocrine systems functioning as they did in our 20's and 30's, our musculature and bone mass strong and supple.

Many of the mental and physical problems that we endure are caused by chronic invisible inflammation. We can't see it, we can't feel it but it goes on day after day, robbing us of our health and our youth.

It was in the year 2000 that I introduced my Inflammation-Aging theory, placing chronic invisible inflammation at the center of age-related diseases and degenerative conditions. The theory was dismissed at the time with ridicule or skepticism—sometimes both! Today, however, mainstream science recognizes the reality of cellular inflammation and its serious threat to health and longevity. In fact, inflammation is the 'new' buzz word.

Because my field is dermatology, where signs of aging and disease are so very visible, I have made it my life's work to intervene; to halt this inflammation; reversing its negative effects, internally and externally.

During medical school and my three-year residency in dermatology, I made important connections between inflammation and disease. To learn about hundreds of skin diseases we studied in books, we also needed to recognize them in clinical examination and under a microscope.

When we examine inflammation under the microscope, it has an unmistakable appearance. To make it visible, we stain the slide. The inflammation shows as dark blue dots, like confetti—although the presence of inflammation is nothing to celebrate.

Quite the opposite.

This "confetti" is also present when we look at aging skin. I was puzzled about the twin occurrence. Could inflammation be causing these changes? I began to consider wrinkles as a *disease*, since inflammation was present when damage to skin tissue resulted in wrinkles.

Whenever I looked under the microscope at everything from arthritis to heart disease, inflammation was *always* a component. Every disease I studied had a common theme, whether it was cancer or aging, inflammation was present.

I was convinced this was not a secondary response. I believed inflammation to be the key to the whole process of disease of every type.

This led me to develop an inflammation-aging theory that is the basis of decades of my research.

Thus began the quest for safe, effective anti-inflammatories that could stop, treat, and reverse the symptoms – without doing harm.

My first startling revelation came with the understanding of the role of diet and nutrition in creating either a pro-inflammatory or anti-inflammatory condition in the body. I learned that the wrong foods will quickly rob us of our youth and health.

In fact, the wrong foods are responsible for rapid, premature aging, a tired, drawn, and doughy complexion, flaccid, weak muscle tone, wrinkled, leathery, dry-looking facial skin, fatigue and poor brain power.

When we eat foods that generate a strong inflammatory response in the body we are actually creating inflammation on a cellular level. The key to the anti-inflammatory diet is that it has been designed to prevent a rapid rise in blood sugar. This is important because a rapid rise in blood sugar causes an insulin response, which then causes an inflammatory response. Our diet can either be pro-inflammatory or anti-inflammatory. Therefore, understanding our food choices is vital if we want to lower inflammation, and it is our single most important step.

There is much else I discovered on the causes and prevention of inflammation that is discussed in more detail in my various books.

But I have taken a biographical detour to make a particular point: new ideas and paradigms are not welcome even in science! This will no doubt be true of the reception awaiting the present book. But we cannot make progress in any scientific inquiry if we do not take the risk of rejection.

This work seeks to show that any study of the Shroud that does not take into account its nature as an organic artifact is ipso facto invalid. This stricture is particularly applicable to the 1988 radiocarbon study conducted by physical scientists without the participation of any microbiologists.

In the course of the book, Steve Mattingly, a microbiologist of national stature, describes his replication of the cleaning procedures used for the 1988 study as applied to a microbe-laden cloth. This replication (repeatable in turn by any scientist) showed that the procedures used would not have eliminated the kind of microbiota present then and now on the Shroud. In short, the chronological dating of the 1988 study was wholly inaccurate because it simply dated the microbiological constituents accumulated over centuries and not the primordial Shroud *per se*.

Microbiology – with particular reference to the skin microbiome and the impact of inflammation – goes further than this. It addresses the enigma of the ages by showing us – at a natural level – how the image on the Shroud was formed and why it remains intact to this day.

For more details read on!

Prologue
The Fifth Gospel

Imagine, if you will, that you had an opportunity to see Jesus as he truly looked on the day of his crucifixion: a real-time image of Jesus on the most important date in history: the day of his proclaimed redemption of humanity. Imagine further that this real-time image by its very nature also bears witness to the resurrection of Jesus, his rising from the dead. And imagine too that this image was formed from the very flesh and blood of Jesus – from the Jesus microbiome, his skin bacteria (*Staphylococcus epidermidis*), and from the blood that poured out of his wounds. Such a real-time image would be what some have called a fifth gospel. It would be a living embodiment of the textual narratives of the Gospel. A tangible testimony to the suffering, death and resurrection of Jesus and of his infinite love for the human race. Such an image, if it existed, would be, outside the Bible, the greatest treasure of Christendom, the most precious possession of the patrimony of humanity.

We try to show in this book that the Shroud of Turin, the purported burial cloth of Jesus, is indeed such a treasure: an image formed from and by the Jesus microbiome. That it is a visual, real-time manifestation of the most important events recorded in the Gospel narratives: the world's ONLY palpable connection to the scourging, crucifixion, piercing and death of Jesus. It is the fifth gospel. It is an archaeological and microbiological witness to the Passion narratives. As the burial sheet of Christ, it shows believers the literal face of their Savior. It is also a witness to the Resurrection simply because it is an *empty* shroud: the body has disappeared. If the body had been moved, this movement would have been reflected on the Shroud. If it had decayed, the Shroud would have decayed as well. The Shroud, then, is an Instagram from the first century: photo-transmitting for future generations the greatest story ever told.

In recent years, however, the Shroud itself has been laid in a shroud as both the media and the general public accepted the verdict of the 1988 radiocarbon dating investigators who concluded that the artifact dated back only to the Middle Ages. Using demonstrable and replicable procedures, however, the present book definitively shows that this carbon dating study, although a testimony to human ingenuity, did not de-contaminate the Shroud at a microbial level and so its dating conclusions are invalid. In fact, by the very nature of the case, as we will see, it will *never* be possible to accurately *carbon date* the Shroud. Thus a new carbon-dating study will not be helpful.

The book also goes further in answering the central mystery of the Shroud of Turin: how was the image formed? Just as the invention of photography first enabled us to see the image on the Shroud in all its glory in 1898, so also 21st century microbiology tells us how the image was created and preserved. By demonstrating the authenticity and antiquity of the Shroud, the present book will complement the faith of millions while also enabling skeptics to look at the four Gospels in the light of the Fifth Gospel.

Introduction
From Relativity and Quantum Theory to the Shroud of Turin

How New Data Creates a New Order

Microbiome = the body of microorganisms that inhabit a specific environment: "the totality of microorganisms and their collective genetic material present in or on the human body or in another environment."

(Dictionary.com)

At the end of the 19th century, leading physicists announced the end of physics noting that only two puzzles remained to be solved: the Michaelson-Morley experiment's failure to show the existence of the ether, a substance that supposedly filled space, and the phenomenon of black-body radiation. Amazingly enough, it was precisely these two seemingly trivial problems that gave birth to the golden age of modern physics with the theories of relativity and quantum mechanics, discoveries that changed our whole understanding of physical reality.

In a striking parallel, over three decades ago researchers announced the "end" of the Shroud of Turin, the cloth revered as the burial garment of Jesus Christ. They had deployed the latest dating technology of the time (AMS Carbon 14 radiocarbon dating) in an attempt to determine the Shroud's true age. The verdict from this study was as lethal as the fiery flames that often enveloped the Shroud's various homes (as, for instance, in 1532 and 1997). The material under scrutiny, and therefore the Shroud as a whole, said the investigative team, dated back only to the Middle Ages. Science had spoken. The case was closed. And the Shroud itself was interred in the shroud reserved for debunked superstitions.

Until now. *Until "the microbial century."* As with the purported end of physics, a few nagging puzzles had remained after the shrouding of the Shroud: How was its image formed and preserved? Why did its material show traces of pollen from the Middle East? What about the reports of a shroud with the image of Jesus going back to the second century? Or the pre-medieval public displays of an image of Jesus not made with "human hands"? Some believers responded to skeptics by offering supernatural explanations for the formation of the image. But the "supernatural" cannot explain away what is known at a natural level. And natural methods, in particular the C14 dating study, seemed to show that the Shroud was not as old as traditionally believed. Yet the puzzles remained.

Until now. *Until this century's paradigm-shifting scientific breakthroughs unveiled a whole new understanding of the image on the Shroud.* Very much like entirely novel modes of scientific thought a hundred years ago gave us a revolutionary perspective on the very large – relativity theory – and the very small – quantum physics. While genetic discoveries dominated the 20th century, microbes "own" the 21st (take, for example, the 2008 Human Microbiome Project). And microbiology has now solved the mystery that had alternatively baffled and misled a Pandora's box of specialists by opening a window into the first century. *We now KNOW the true story of the Shroud.*

The 1988 Carbon 14 dating investigation was demonstrably flawed, fatally so, because it failed to consider the elephant in the room, namely the most ubiquitous life-form in the universe – the trillions of microorganisms that pervade our planet and transform their immediate environments. This was due to no fault of the investigators who dispassionately applied a methodology that has more often than not been effective. But microbiology, the science most relevant to an analysis of any organic artifact like the Shroud with its *embedded microbiome*, was not considered in the course of the study. There were no microbiologists on the investigative team. Consequently (and understandably) there was no effort to analyze the scientifically indisputable part played by known microorganisms, in fact tens of thousands of kinds of microorganisms, in the history and therefore date-ability of a burial cloth such as the Shroud.

Until now. *Until the paradigm of the microbiome was brought to the table.* At the invitation of Shroud researchers around the turn of the century, the microbiologist co-author of this book, Stephen J. Mattingly (SJM), who was President of the Texas Branch of the American Society for Microbiology and a Professor of Microbiology and Immunology at the University of Texas Health Science Center at San Antonio for 33 years, launched his own investigation. The results of this study have been as dramatic as the 1898 photograph of the Shroud that showed its image to be a photographic negative. If the findings which guided studies of the Shroud in the 20th century might be called Shroud 2.0, the insights of microbiology have given us Shroud 3.0 for the 21st century.

By re-examining the persisting post-Carbon 14 dating puzzles in the light of microbiology, SJM *replicated* the processes that led to the formation and preservation of the Shroud image. His study, carried out in cooperation with the co-inventor of the AMS Carbon 14 methodology and presented to a national meeting of the American Society for Microbiology, yielded *instantly and universally verifiable* claims and predictions shared here for the first time. And the biometric verdict *on a natural level* is that the Shroud, via its embedded microbiome, is the first photograph in human history – a photograph enabled in large part by the very microorganisms that were not even considered in prior investigations. It is, therefore, as the religious believers of the first millennium called it, an image "not made with human hands."

Overview – *The Shroud*

Attributes	Straw yellow-colored image of crucifixion victim that acts like a photographic negative. If photographed, gives positive image. Negative of positive image gives more defined image. 3D structure. Photographic features discovered only with 19th century invention of photography. 3D attributes seen only with the aid of 20th century technologies.
Image Origin	Bleeding of the Man of the Shroud led to an explosion of skin bacteria (e.g., *Staphylococcus epidermidis*). Unsaturated fatty acids from the skin cells and bacteria oxidized when the Shroud was exposed thus forming the image. This process is replicable and was understood only with the advent of 21st century microbiology and the study of microbiomes.
Preservation	The growth of innumerable other microorganisms over time stabilized the surface and interior regions of the linen fabric holding it together making the Shroud a self-repairing artifact. *Dating the Shroud using conventional methods such as radiocarbon dating is not possible – and will never be possible – because the bacterial and fungal cell walls of microorganisms colonizing the Shroud over the centuries have the same chemical structure as the cellulose that is the primary component of linen.* Dating the Shroud simply means dating its latest microbial inhabitants.
Image and Blood	Major medical authorities and medical examiners who have performed tens of thousands of autopsies have studied the bloodstains and the wound images on the Shroud and have concluded that the Man of the Shroud was scourged, crucified, crowned with thorns and pierced with a lance as described in the New Testament Gospel narratives. Blood chemists found human blood on the Shroud.

Externalities	Botanists have found pollen grains on the Shroud from first century Jerusalem. Chemists detected dust particles usually found in the tombs of Jerusalem. Textile experts found that the style of stitching on the cloth was also found in the first century ruins of Masada in Israel.
Complement	The facial and neck stains, nose structure and blood group on the Shroud match those on the Sudarium of Oviedo (in Spain) that is believed to be the face cloth that covered the head of Jesus prior to his burial.
Mystery	What happened to the Shroud's body? Shrouds decompose with their bodies. Studies of the image show the body of the Man of the Shroud did not decompose and was not moved. It appears to have dematerialized.
History	The Gospels describe the empty Shroud. In the first to the fifth centuries there were scattered reports of the Shroud of Jesus being preserved. From 400 A.D., the Image of Edessa became known as an image of Jesus not made with human hands. Shortly after there was an explosion of icons of Jesus that resemble the face on the Shroud. The Image was moved to Constantinople in 944. Visitors to Constantinople reported seeing the Shroud of Jesus. The Pray manuscript shows that the image on the Constantinople shroud resembles the Shroud of Turin image. French Crusaders sacked Constantinople in the thirteenth century and stole many of its relics. The Shroud re-appeared in France in the fourteenth century.

Section I
Shroud 3.0

1.1 The Once and Future Shroud

Hic iacet Arthurus, rex quondam, rexque futurus
"Here lies Arthur, king once, and king to be."
　　　　　　　　Sir Thomas Malory, *Le Morte d'Arthur* 21:7:

The New Testament Gospels are narratives of the life of Jesus of Nazareth. The passion, death and resurrection of Jesus play a pivotal role in these accounts. Some have said that this textual record from the first century does not stand alone. It is complemented by a visual portrait. A live image. The burial garment of Jesus. The Shroud of Turin.

The Shroud, a 14.5 by 3.7 foot linen cloth with a three-to-one weaving pattern called "herringbone twill," is a unique burial artifact because imprinted on it in a seemingly inexplicable fashion is the bloody image of a crucified man. In fact, an image of someone who seems to have mysteriously dematerialized through the Shroud. No other burial cloth in history has left us with the image of the person buried in it – and, even more astounding, an image that indicates the disappearance of that person. For centuries, the Shroud was revered as the cloth in which the body of Jesus was wrapped after the crucifixion. But with the invention of photography, the barely visible visage on the Shroud was revealed to be an image without parallel in history.

Normally, after a photograph is taken, a negative, a reversed order image (e.g., light area appears dark and vice versa, left and right are reversed) forms on a plastic film or glass or other material as it is developed. The negative is then converted into a positive print on special paper through a series of steps including a second reversal. The Italian Secundo Pia took the first photographs of the Shroud in 1898. To his astonishment, the image on the Shroud developed as a positive instead of a negative (with one exception: the bloodstains developed as negatives). The image as a whole was a photographic negative – although it predated the invention of photography by hundreds of years! Moreover, what was previously not visible to the naked eye was now seen to be a real-time image of a man scourged and crucified exactly as depicted in the Gospel Passion narratives.

Subsequently, advanced 20th century observational technologies, from direct microscopy and infrared spectrometry to X-ray radiography, thermography and ultra-violet fluorescence spectrometry revealed previously undiscovered features of the mysterious image, especially its three-dimensional nature. Whereas a regular photograph or painting is flat and gives us two-dimensional data on its image, a statue gives us three-dimensional data given that it has height, width and depth. The Shroud is somewhere in between a picture and a statue because it carries data on the parts of the body around which it was wrapped and its distance from these parts. If it were wrapped around the 3D body, it would assume the shape of the body without distortion but "straightens" out if taken off.

But, despite its extraordinary features, time after time the Shroud has been unceremoniously laid to rest in the modern era. Amazingly, each time it rose again, restored to life by the very scientific method deployed against it.

Right after it was first photographed in 1898, distortions caused by the image transmission technologies of the day led to widely believed accusations that the cameraman had doctored the photographs. Fortunately, efforts to show the photographs to be genuine eventually began to bear fruit. Just then, a French priest published an influential article in 1900 arguing (on mendacious grounds) that the Shroud was a medieval painting. Yet, shortly after, a renowned French scientist directed a research study demonstrating that it could not be a painting. A report on this study was delivered to the French Academy, the most prestigious scientific body in France, in 1902, where it was enthusiastically received. But an atheist serving as secretary of the Academy blocked a vote to ratify the report. Consequently, the initially friendly media swiftly turned hostile. Catholic Church authorities, unnerved by the opposition, kept the Shroud out of sight until a public exposition in 1931.

Fierce arguments for and against the Shroud's authenticity continued after the exposition. Then, in 1978, Church authorities allowed a team of 35 multi-disciplinary scientists constituting STURP, the Shroud of Turin Research Project, to spend five full days subjecting the Shroud to intensive scientific scrutiny. They used a variety of instruments and methods. The STURP team came from nearly 20 corporations and research institutions ranging from IBM and Lockheed to the Jet Propulsion Laboratory and Los Alamos National Scientific Laboratories. Members included agnostics, Jews and Christians. This was the first and only hands-on study done on the Shroud in its entirety and its conclusions were positive in nature.

This is the official STURP summary of its conclusions:

No pigments, paints, dyes or stains have been found on the fibrils. X-ray, fluorescence and microchemistry on the fibrils preclude the possibility of paint being used as a method for creating the image. Ultra Violet and infrared evaluation confirm these studies. Computer image enhancement and analysis by a device known as a VP-8 image analyzer show that the image has unique, three-dimensional information encoded in it. Microchemical evaluation has indicated no evidence of any spices, oils, or any biochemicals known to be produced by the body in life or in death. It is clear that there has been a direct contact of the Shroud with a body, which explains certain features such as scourge marks, as well as the blood. However, while this type of contact might explain some of the features of the torso, it is totally incapable of explaining the image of the face with the high resolution that has been amply demonstrated by photography.

The basic problem from a scientific point of view is that some explanations which might be tenable from a chemical point of view, are precluded by physics. Contrariwise, certain physical explanations which may be attractive are completely precluded by the chemistry. For an adequate explanation for the image of the Shroud, one must have an explanation which is scientifically sound, from a physical, chemical, biological and medical viewpoint. At the present, this type of solution does not appear to be obtainable by the best efforts of the members of the Shroud Team. Furthermore, experiments in physics and chemistry with old linen have failed to reproduce adequately the phenomenon presented by the Shroud of Turin. The scientific consensus is that the image was produced by something which resulted in oxidation, dehydration and conjugation of the polysaccharide structure of the microfibrils of the linen itself. Such changes can be duplicated in the laboratory by certain chemical and physical processes. A similar type of change in linen can be obtained by sulfuric acid or heat. However, there are no chemical or physical methods known which can account for the totality of the image, nor can any combination of physical, chemical, biological or medical circumstances explain the image adequately.

Thus, the answer to the question of how the image was produced or what produced the image remains, now, as it has in the past, a mystery.

We can conclude for now that the Shroud image is that of a real human form of a scourged, crucified man. It is not the product of an artist. The blood stains are composed of hemoglobin and also give a positive test for serum albumin. The image is an ongoing mystery and until further chemical studies are made, perhaps by this group of scientists, or perhaps by some scientists in the future, the problem remains unsolved.[1]

But what science gave with one hand, it seemed to take away with the other. Not science as a tool for immediate observation but science as a tool for inference. STURP was followed ten years later by the fateful radiocarbon analysis. Scientists conducting this dating study concluded that "the linen of the Shroud of Turin is mediaeval." Overnight the artifact that seemed to visually confirm what was textually recorded in the Gospels became an object of scorn. It was of relatively recent not ancient origin, said scholars and commentators. A fake and a forgery not a fulcrum of faith. Shroud-ers, sniffed one British academic, were "flat-earthers."

Here the matter may have ended with Shroud scholarship reduced to a Charge of the Light Brigade. Except that the discrediting study was itself discredited by the discovery of new evidence buttressed by fresh insights into the body of already established hard data.

For instance, in 2005, Philip Ball, the editor of the world's most prestigious scientific journal, *Nature*, which had indeed published the 1988 Carbon-dating report, said in the same journal: "Perhaps more compelling is that most of the shroud lacks vanillin, a breakdown product of the lignin in cotton fibres. There is vanillin in the Holland cloth [the Shroud's backing cloth added in the Middle Ages], and in other medieval linen. Because it decomposes over time, this suggests that the main body of the cloth is considerably older than these patches. By calculating the rate of decay, [Raymond] Rogers arrives at his revised estimate of the shroud's age. ... It is, he says, between 1,300 and 3,000 years old. Let's call it somewhere around the middle of that range, which puts the age at about 2,000 years. Which can mean only one thing..."[2]

The *Nature* article also drew attention to co-author SJM's work seeking to show, as Ball put it, "that bacteria and fungi on the fibres had skewed the dates, by a thousand years or so." The present book tells "the rest of the story" by exploring the bio-history of the Shroud with a focus on the nature and origin of the image etched on it. Furthermore, it conclusively shows that the decontamination procedures adopted by the 1988 investigators did not remove the Shroud's microbial inhabitants – thus negating the evidential value of their analysis. At best, the dating exercise would have yielded an overall "age" of all the Shroud's living and dead microbes.

With these latest findings, the Shroud rises yet again, emerging from the cloud of suspicion cast by the pre-microbial-era studies. The Secondo Pia photographs launched Shroud 2.0, the scientific study of the artifact in the 20th century. The matrix of microbiology has now mapped out a radically novel and hitherto unexplored horizon in

Shroud research thus launching the era of Shroud 3.0. Fittingly, this breakthrough has come to fruition in the 21st century, "the microbial century."

Our inquiry into the Shroud's bio-history will be complemented by other available evidence from ongoing scientific and historical studies. We will rely on and reference only data generated by specialists in the relevant disciplines. Our goal is to see the big picture:

- What do we have "before us" on the Shroud?
- How did the image embedded in it come to be?
- How was the Shroud physically preserved?
- What do we know of its history?
- Who is the Man of the Shroud?
- Why are some thinkers skeptical about the traditional account of the Shroud's origin?
- Is there any deeper meaning to the Shroud, any "moral" of the story?

As is evident from the terms of the scope outlined here, our inquiry seeks to be holistic and comprehensive. Consequently, it should be of interest to a diverse audience: the faithful, of course, but also the scientific community and the world at large. The Shroud with its unique image is an undeniable datum calling out for an explanation: even after the negative results from the 1988 radiocarbon dating, mainstream scientists admitted that they had no idea how the image was created.

As Ball wrote in his *Nature* article: "The shroud is a remarkable artefact, one of the few religious relics to have a justifiably mythical status. It is simply not known how the ghostly image of a serene, bearded man was made. It does not seem to have been painted, at least with any known historical pigments." Esoteric writers have proposed theories built on outlandish speculation – it was a Da Vinci painting or the product of a hitherto-unknown medieval proto-photography technology – but no such account has won scientific acceptance. Numerous such pseudo-theories on the formation of the image have been proposed and, after the usual half-life of sensational sound-bites, promptly put out of their misery.

Of course, given its fame, the Shroud has inspired numerous works. Nearly two thousand books have been written on the Shroud. So what more needs to be said and why should anyone care?

The answer is that the mystery remains. If a plausible resolution to the mystery is to be possible, it has to:

- come from science;
- be compatible with what is already known scientifically about the Shroud including the Carbon 14 results;
- go beyond current scientific paradigms, which have failed to provide a path forward, while being grounded in demonstrable hard facts;
- originate from a specialist in the specific area of discussion;
- comport with what is known historically about the Shroud.

The present work meets all these criteria. It is:

- Restricted to the natural, which in this instance is the purview of science, with no appeal to the supernatural in "explaining" the origin and nature of the image;
- Entirely compatible with the Carbon 14 data (although not the popular interpretations of this body of data which inaccurately presume decontamination) as well as various other specialist-generated scientific data;
- Introducing an entirely new paradigm in exploring the question using mainstream microbiology, an obviously relevant discipline ignored by previous researchers, and applying its latest methodologies;
- Co-authored by a veteran microbiologist who has studied the Shroud from the perspective of his discipline for over two decades in dialogue with prominent Shroud researchers representing a variety of viewpoints;

- Continuous with what is known about the various "sightings" of the Shroud in history and various accounts of it across the centuries.

On this last point, we must be aware that an appraisal of the pre- and post-medieval history of the Shroud will necessarily be unlike the study of any other artifact. This is because the kind of artifact we today know the Shroud to be (a photographic negative in 3D produced by a microbiome) should guide the way in which we consider its provenance and past.

An analogy might help. Let's say a construction company in a remote corner of Siberia discovers a rock that appears to defy the known laws of motion while also emitting a hitherto unknown form of electromagnetic radiation. The press and the public conclude that this rock must have an extra-terrestrial origin. But scientists studying the rock dismiss the extra-terrestrial theory as a superstition because the rock shows no sign of having arrived from anywhere else. How the rock manifests apparently extraordinary attributes and how it came to be located where it is, they say, is still unknown but later advances in science will eventually produce a suitable theory of origin. At this point, it can be safely said that the rock is part of an elaborate hoax. The rock's unique attributes are irrelevant because they simply show the ingenuity of the hoaxer and further studies will eventually replicate the attributes. That it is a hoax is certain because no object can "violate" the known laws of nature. After all it is just a rock. Academic historians analyzing the rock say its existence was never formally reported or recorded in bygone eras and it should therefore be considered a "plant" from some hoaxer. That it is a hoax is clear because there is no buttoned-down history of how the rock got here.

A minority group of scientists, however, challenge their colleagues' view. They hold that we should take the rock on its own terms. Its properties do not "violate" the laws of nature. Rather they indicate that we might be dealing with something that came to us from another galaxy or from a "Fermi bubble" in our own galaxy and there may be a different class of laws governing particles from such a realm. We should study the rock with an eye to understanding these other laws instead of shutting down any discussion of this possibility: and keep in mind we are still operating within the realm of the natural. Of course, this means that the rock would have landed on earth from a distant source. Whether or not that happened is a matter of history. But this is where academic historians point to the lack of data confirming such an event.

Here the minority scientists are joined by a minority group of historians who hold that you cannot study the history of the rock without taking into account its self-evidently unique nature. In doing so, they recognize that the only way to trace the prehistory of the rock is to track down suggestive clues rather than to insist on the kind of definitive data characteristic only of modern history. After all, archaeologists, anthropologists and historians, as a matter of course, plausibly reconstruct past societies and events from the barest fragments of physical hard facts. The minority historians then point to the hazy legends of native Siberians and disjointed reports from ancient nomads that talk of a strange rock with ethereal properties. There are even centuries-old drawings of a rock landing in a Siberian field in a blaze of light. While none of this constitutes foolproof evidence of the rock's extra- or intra-galactic origin, we now have a plausible picture that suggests this to be the case. It is the very nature of the rock – its attribute structure – that drives the discussion and enables historical re-construction scenarios.

This analogy provides a path to navigate through debates about the Shroud's history. The unique nature of this artifact is a Geiger counter for detecting "sightings" of the Shroud over the centuries. These sightings coupled with the extraordinary nature of the artifact show us that the Shroud has been with us some 2000 years.

With this we begin our inquiry into the Shroud. The first step is an overview of ten hard facts established by Shroud 2.0 in (for the most part) the twentieth century.

1.2 Shroud 2.0 Hard Facts

The scientific study of the Shroud began with Secondo Pia's 1898 photographs. Over the next century, inter-disciplinary specialists made successive contributions to the investigation. Their body of breakthroughs constitutes Shroud 2.0. The heyday of Shroud 2.0 lasted from 1898 to 1988 when serious research was virtually brought to a halt by the red STOP light of the radiocarbon study. The red light, as we will show, was a red herring. But even before the microbiome paradigm rendered carbon-dating irrelevant, Shroud 2.0 laid out ten enduring hard facts pointing to the conclusion that the Shroud of Turin was the burial cloth of Jesus.

The single most significant hard fact is the Shroud itself: staring right in our faces is the facticity of its mystery: a photographic negative created before the invention of photography; an image that portrays in excruciating detail the tortures inflicted on the Gospel Jesus; bloodstains at the precise points where the narratives place them; pollen grains from first century Jerusalem and sand particles from the reputed location of Jesus' tomb. This – the artifact as we have it – is the fundamental hard fact that must be taken as the starting-point of any serious investigation of the Shroud.

Shroud 2.0 studies started with this hard fact. Over several decades, hundreds of scientists and other scholars assiduously assembled bits and pieces of relevant data that grew into a time-tested unified field theory. In terms of the consistency of the internal evidence, STURP seemed like the last word testifying to the authenticity of the Shroud. In terms of the Shroud's antiquity, there were two challenges: its public history and the dating of its material embodiment. What could be known of its initial history was subject to the vicissitudes of history itself particularly in the first millennium after Christ. But evidence of the antiquity of the Dead Sea Scrolls did not depend on evidence concerning its "public" history: likewise with the Shroud. As for dating the physical material of the Shroud, various approaches were possible, radiocarbon dating being just one. The central problem facing dating analyses was contamination over the centuries. Accurate dating was not possible without true decontamination and this turned out to be the Achilles Heel of the C14 dating.

But before reviewing the question of dating, consider the ten hard facts generated by Shroud 2.0:

Unique Characteristics of the Shroud Image.

In a paper published in the journal *Science*, Barbara Culliton described the extraordinary nature of the Shroud image: "What astonished [Secondo] Pia, and continues to astonish Shroud scholars, is that the image that appeared on his photographic plate was not a characteristic negative ... Instead, Pia's negative showed all the qualities of a positive print. The image of the Man of the Shroud showed gradations of tone that gave the body depth and contour. The face had the qualities of a photographic likeness, not the flatness of a negative. Thus, it seems that the Shroud itself must be, or possess some of the properties of, a photographic negative. It is as if the cloth were a piece of film."[1] The STURP study showed that "the image has unique, three-dimensional information encoded in it." The image density conveyed spatial information. The 3D feature was as unexpected as it is unusual: "The cloth-to-body distance correlates so precisely that the image perfectly encapsulates three-dimensional data perfectly. When the shroud image is fed into NASA's VP-8 image analyzer, it produces a bas-relief [sculpture] of the man of the shroud with no distortion. No other image, drawing, painting, or photograph has this quality—only star maps and the shroud image; everything else distorts."[2] Further, the image is not painted on the linen but is embedded in its very structure.

The Man of the Shroud and the Gospel Passion Narratives

The image on the Shroud matches the profile of the subject of the Passion narratives in the New Testament Gospels. Jesus is said to have been scourged, crowned with thorns, crucified and speared. The wounds and blood flow patterns of the Man of the Shroud precisely reflect these specific actions: some one hundred whip marks; wounds on the wrists, feet, hands, side and head; and bruises on the face, knees and shoulders. Former Medical Examiners, who have conducted thousands of autopsies, and other medical doctors have confirmed the correspondence. Two features stand out: the crucifixion wounds appear on the wrists and not the palms and there are head wounds consistent with the crowning. All medieval crucifixion paintings show the nails piercing the palms. Only in later centuries was it discovered that the Roman method of crucifixion could also entail driving nails through the wrists and not just the palms. So this was no medieval painting! Also, although thousands of victims were crucified by the Romans and others, none were crowned with thorns other than the Jesus of the Gospels and the Man on the Shroud.

Bloodstains on the Shroud

Prominent chemists such as John Heller, professor of Internal Medicine and Medical Physics at Yale University, and blood specialist Alan Adler (who was Jewish) performed multiple chemical and immunological tests to show that the bloodstains on the Shroud had the chemical characteristics of blood: hemoglobin, serum albumin, etc.. Their studies were published in peer-reviewed scientific journals. One dramatic instance was their paper "A comprehensive examination of the various stains and images on the Shroud of Turin" which appeared in *Archaeological Chemistry*, a journal of the American Chemical Society. It should be noted that the Shroud also contains traces of iron oxide but this is to be expected given a number of factors ranging from the linen retting process to the iron in the bloodstains to the fact that, in previous centuries, paintings were touched to the artifact so as to sanctify them.

Botanical Profile

Botanical experts have detected pollen grains on the Shroud that are found in the Israel-Palestine region. Three major studies in well-known peer reviewed journals, *Nature* (2015), *Archaeological Discovery* (2015) and *Archaeometry* (2016), indicate a Palestinian origin for the pollen in the Shroud. Professor Avinoam Danin (yet another Jewish Shroud researcher), a world authority on Middle Eastern flora from Israel's Hebrew University, found a profusion of pollen grains from Israel on the Shroud. In Danin's view, the Shroud was an ancient Israelite cloth.

Dust

The body of Jesus was laid to rest in a cave. Researchers have detected traces of limestone dust on the Shroud. Chemical analysis of the dust showed that it is similar to the limestone dust found in other ancient tombs in Jerusalem. "Joseph Kohlbeck and Richard Levi found travertine aragonite limestone particles in sample dust collected from the Shroud's surface. Kohlbeck was Resident Scientist at the Hercules Aerospace Center in Utah. Levi-Setti was with the Enrico Fermi Institute at the University of Chicago, Using a high-resolution microprobe, Kolbeck and Levi-Setti compared the spectra of the dust from the Shroud with samples of limestone collected from ancient tombs in Jerusalem. They found that chemical spectral data were identical except for some minor bits of cellulous fiber that could not be removed from the dust."[3]

Documentary and Circumstantial Evidence from the First Millennium

Three ideas, unique in human history, arose soon after the end of Jesus' ministry. A focus on the shroud left behind in Jesus' tomb; the claim that there was an image of Jesus not made of human hands; and the appearance of the Shroud of Jesus in Constantinople. These three ideas serve as a template for an exploration of the history of the Shroud.

This history may be classified under pre- and post-medieval eras and further divided into the ancient, Byzantine and modern-medieval phases. Its history in pre-medieval times is speculative given the state of persecution and war of that era. For the first five centuries of Christianity, its adherents communicated in codes and symbols for fear of their enemies under what is called "The Discipline of the Secret." Nevertheless, there are unmistakable references to the Shroud's existence: "the Lord, after having given the Shroud to the servant of the priest" (second century), "the portrait of Jesus" (*Doctrine of Addai*, 400 A.D.), "the divinely wrought image which hands did not form" (590 A.D.), "Peter ran with John to the tomb and saw the recent imprints" (seventh century).

The Image of Edessa, an image of Jesus reportedly not made by human hands, has been written about at least since 400 A.D. In 944, the Image was moved to Constantinople as narrated in histories of the time: the "image not made by man of Christ our God was transferred to Constantinople." It was also called the Mandylion and was described as "an impression of God's assumed human form by a moist secretion without coloring or painter's art. An impression of the face was made in the linen cloth."[4]

Constantinople, the reliquary of Christendom, claimed to possess the burial cloth of Christ and contemporary reports indicate the presence there of an artifact resembling the Shroud. The Hungarian Pray Manuscript, dating back to 1192 (and therefore pre-dating the Sack of Constantinople), has an illustration of the burial cloth of Jesus that looks very much like the Shroud of Turin. Art, devotional practices and historical reports collectively seem to indicate that the Image of Edessa=the Shroud of Constantinople=the Shroud of Turin. In times of turmoil, there is not much beyond this we can hope for in terms of recorded history. Since most mainstream historians have accepted the results of the 1988 carbon-dating, they have not seriously considered the Shroud's history in the first millennium. This is a good example of how an error in one discipline can distort studies in another.

Public History from Medieval Times

Constantinople was conquered by the Crusaders in 1204 and the Mandylion/Shroud disappeared at that time. One of the Crusaders wrote: "There was another of the churches ... where was kept the syndoine in which our Lord had been wrapped, which stood straight every Friday so that the figure of our Lord could be plainly seen. No one, either Greek or French even knew what became of the syndoine after the capture of the city."[5] But another writer reported in 1205 to the then Pope that the "predators" preserved "the sacred linen in Athens." Athens had been taken over by the French Crusaders in 1205. In 1357, a French knight, Geoffrey II de Charny, displayed the Shroud as we know it today in a church in France. The public history of the Shroud begins from this point. His descendants passed it on to the House of Savoy which became the ruling family of Italy. The family bequeathed the Shroud to the Catholic Church in 1983.

The Sudarium of Oveido

The Shroud is complemented by what is called the Sudarium of Oveido, a blood-stained cloth that is believed to have been placed on the face of Jesus after the crucifixion. The Gospel of John tells us that there was a headcloth: "When Simon Peter arrived after him, he went into the tomb and saw the burial cloths there, and the cloth that had covered his head, not with the burial cloths but rolled up in a separate place." (John 20:6-7). It was a common practice among the Jews to place a cloth on the face of a deceased person while the body was being prepared for burial. The face cloth was removed before the corpse was wrapped in linen. The Sudarium was believed to have been in Palestine until 614. After the Persians conquered Jerusalem it was transported to Alexandria until finally reaching Spain. The Sudarium shows only bloodstains and no image. Its blood type matches the blood type on the Shroud (AB) and its facial and neck stains match the stains on the Shroud.

Shroud Image as Icon Template

The icons of Jesus that emerged at least from 550 A.D. reflect the image of the Man on the Shroud. There can therefore be little doubt that this image served as the template for the various icons of Jesus which means the Shroud was known to Christendom since its earliest years.

Resurrection

On a natural level, the Shroud also bears witness to the Resurrection. We know that the crucified body of Jesus was wrapped in a linen shroud: "Taking the body, Joseph wrapped it [in] clean linen and laid it in his new tomb that he had hewn in the rock." (*Matthew* 27:59-60). The Gospels tell us what the first witnesses saw: "But Peter got up and ran to the tomb, bent down, and saw the burial cloths alone." (*Luke* 24:12). So we see here a shroud without a body. But what is remarkable is the fact that the Shroud survived. Burial shrouds normally do not survive because when the body decomposes, its shroud decays with it. Bacteria and fungi in the soil contain enzymes that decompose the body and its shroud. Cellulases specifically break down cellulose, the major structural component of linen. And arrays of enzymes are responsible for the decomposition of non-skeletal human tissues. But if the body in the Shroud did not decompose, if it "left", then the Shroud as we have it would be what we should expect. Paradoxically, the Shroud was preserved because its body had disappeared. Moreover, it is clear that the body was not physically removed because the blood clots seen on the Shroud were not smeared or broken which they would have been if the body was moved. Finally, the Shroud shows that its body suffered no decomposition which indicates that the body "disappeared" soon after it was left in the tomb.

So what is the significance of these ten facts? When we connect these dots we get a picture – a real-time image of Jesus. Science begins with a hypothesis to explain a set of data after which it aggregates and analyzes supporting evidence. The Shroud furnishes us with both evidence and explanation right at the start as our starting point. The "explanation" handed down centuries ago is that this is the burial cloth of Jesus. The "evidence" supporting the explanation became increasingly visible (literally!) over the centuries and especially in the modern era. And today microbiology has verified not only the theses of Shroud 2.0 but the proclamation of the first millennium: the Shroud of Turin is "the image not made by man of Christ." This is the primary breakthrough of Shroud 3.0.

But the breakthrough makes sense only in the light of the 21st century discovery of the microbiome. We will take a break from Shroud history to explore the amazing world of the microbiome.

1.3 Microbiome 101

Hardly a day goes by without some reference in the various news outlets to the role of microorganisms in our daily lives. It can be the latest information on the spread of a lethal virus to the most recent multiply-antibiotic-resistant bacterium causing disease. But in reality microorganisms have always had the upper hand on all aspects of life on planet earth. It is believed that they emerged as the first life-forms. We recognize today that they directly or indirectly provide all our food. They are partners in photosynthesis producing edible crops and releasing oxygen into the atmosphere. They fertilize our crops producing materials that enrich the soil and detoxify harmful chemicals. Microbes break down plant materials in the stomachs of cattle providing them with energy and food. They are also involved directly in producing foods such as cheeses, butter, bread, and a variety of fermented beverages. They are primary sources for a variety of nutrients and supplements, such as vitamins, amino acids, sugars, fatty acids and other essential nutrients. They are the primary source of most of our antibiotics and many pharmaceuticals. They developed these antibiotics not for us, but to ward off their microbial competitors long before *Homo sapiens sapiens* walked the planet.

Much of the world's oil supplies are due to microbial activity and decomposition of organic matter. Decay of vast amounts of vegetation, large and small animals and microbes themselves through millenia have formed massive deposits of energy-rich organic matter that have escaped the complete breakdown to carbon dioxide by oxidation by being buried deep within the earth's interior. Microbes clean up hazardous solid and liquid wastes both natural and man-made. Microbes are not moral or immoral; they simply exist and utilize their unique lifestyles to survive and reproduce as do all organisms. They were the first organisms to appear on the earth and they will be the last assuming some cataclysmic event eliminates higher forms of life.

The term microorganism refers to the fact that they can only be seen using a microscope which greatly enlarges the cell. Many life forms become multi-cellular and visible to the naked eye, while microbes rarely go through complex life cycles and differentiate into specialized cells, organs and tissues. We first became aware of microbes because they were found to cause diseases, such as anthrax, tuberculosis, among many others. It became possible to associate the specific disease with the microbe seen under the microscope. This was the beginning of the study of infectious disease which has impacted all of our lives to the present time. Human deaths from epidemics of cholera, plague, smallpox, tuberculosis among others have far exceeded deaths caused by all the combined military campaigns throughout the history of humanity.

Once the existence of the microbe was established, it became apparent that these smallest of creatures had a fundamental impact on every aspect of life on planet earth. Although some have referred to microbes as primitive forms of life, their very existence today indicates that they have adapted and survived throughout the history of life. It has been hypothesized that multi-cellular forms of life originated when individual microbes began to associate and develop dependence on one another. An example is mitochondria, which provide energy for plant and animal cells. Mitochondria are the size of bacteria and are believed to have been engulfed by larger microbial cells and both cells benefitted from the association. Chloroplasts are also the size of bacteria and a similar process of engulfment took place leading to photosynthesis in higher plants and the ability to produce sugars and release oxygen into the atmosphere.

Once the importance of the microbe became evident in all aspects of life, their distribution also became apparent. They are everywhere! Microbes are found in the hottest as well as the coldest regions of planet earth. One can begin by considering the very room in which one is reading this book. By taking a sterile moistened cotton swab and sampling every square inch of the floor, ceiling, walls, fixtures and furniture in the room and submitting the swabs for genetic analysis using the latest technology, every sample would come back positive for the presence of not one but multiple microorganisms. In that microbiologists have only been able to culture 1% of the organisms known to be present, it is quite likely that there are microbes in your room that have never been described before. Now think outside your room and consider your entire house and the soil surrounding your house or apartment. How about the air that you are breathing? By now you will be beginning to appreciate that you are living as a guest in a "home" owned and operated by an array of microbes that are beyond comprehension. In fact, you owe your very existence to the "lowly" microbe.

An important question may have become apparent at this point. Do microbes operate completely independently or as a complex group closely associated with each other? Undoubtedly, many microbes can function quite nicely as independent species. Examples might include yeasts in fermentation or *Streptococcus pneumoniae* in a bloodstream infection known as septicemia. However, in the vast majority of situations, microorganisms exists in dynamic, complex communities. These are known as microbiomes and include every microbial member present in the community. This discovery has been made possible by the advancements in biotechnology and genetic research that allow the study of all populations present in the community. These advancements have resulted in the accumulation of massive data banks facilitating the identification and characterization of microorganisms associated with both healthy and diseased humans. This massive undertaking is known as the Human Microbiome Project which is a counterpart to the Human Genome Project. It is estimated that over 100 trillion microorganisms are present on the surface and within various regions of our body.

One prominent area in the Human Microbiome Project is the study of the Gut Microbiome. It has identified bacteria that are linked to overall gastrointestinal and general health. The majority of bacteria are anaerobic and difficult to recover by culture. Molecular DNA techniques are the main approaches. Gut microbiome imbalances have been associated with irritable bowel syndrome, inflammatory bowel disease, diabetes, obesity, cardiovascular disease, celiac disease and other diseases. Monitoring and restoring the normal gut microbiome leads the patient to overall health. These are complex and involved studies and this microbiome approach will revolutionize the study of gut diseases.

Another microbiome under study relates to the skin: the Skin Microbiome project seeks to determine the microorganisms associated with skin health and disease. The skin has multiple ecological niches that can support a wide variety of microbes that have been estimated to number over one thousand species. *Staphylococcus* and *Corynebacterium* spp. are generally associated with moist areas of the skin but environmental factors can greatly influence the relative numbers and distribution. Future studies on the Skin Microbiome will help formulate approaches that lead to skin health over the lifetime of the patient.

These two brief summaries of microbiome projects provide a background into the future direction of fundamental microbiological research. The ability to look at entire populations simultaneously allows us to see how changes in the environment can influence the effects on microbes and how they alter the environment in a positive or negative way.

An understanding of microbiomes is also fundamental when examining treasured historical artifacts, such as the Shroud of Turin. Would you not expect such an ancient linen cloth to be covered with a vast array of microorganisms? Are they simply guests on the cloth or have they utilized their resources to grow and reproduce, invading every square inch of the linen material? Especially important in understanding how the image formed on the Shroud is an examination of the microbiome of human skin. In addition, the nature of the crucifixion of Jesus, as recorded in the Gospels, would drastically alter the environment of the skin allowing some organisms to proliferate while others would be slowed greatly or inhibited.

1.4 The Jesus Microbiome Project

"The issue became one not of whether the sample taken in 1988 was contaminated, but whether the contamination was sufficient to skew the Shroud's dating by thirteen hundred years. As was generally agreed, 60 per cent contamination was required of a cloth genuinely of the first century to appear to be of the fourteenth. But surely the presence of 60 per cent contamination would be very obvious? Not so, according to Professor Mattingly, and to his counterpart at the University of Queensland, the late Dr. Tom Loy. As argued by both, the micro-organisms' transparency could cause them to be almost invisible." Ian Wilson, *The Shroud*[1]

"What is the human microbiome? In many ways the best way to define the microbiome is to begin with the simple question, "What are we?" If you'd answer "human" to this question, you're partially correct... But in fact, only [a percentage] of the cells in our body are human!"[2]

"For a man between 20 and 30 years old, with a weight of about 70 kg (154 pounds) and a height of 170 cm (about 5'7) ... there would be about 39 trillion bacterial cells living among 30 trillion human cells. This gives us a ratio of about 1.3:1 - almost equal parts human to microbe."[3]

We are primarily concerned with the bio-history of the Shroud with a focus on the nature and origin of its image. From the standpoint of microbiology, the image was formed by certain skin bacteria (*Staphylococcus epidermidis*) and the Shroud itself was preserved for centuries by the activity of tens of thousands of other kinds of microorganisms. These two fields of microbial activity constitute the Shroud's embedded microbiome. "Microbiome" here refers to the microorganisms that inhabit a particular environment – in this case, the Shroud – creating their own eco-system. Since the Shroud is believed to be the burial cloth of Jesus, the skin bacteria that created its image derived from Jesus: *in brief, the Shroud image was created by Jesus' microbiome.*

In studying Jesus' microbiome, we are like Crime Scene Investigation (CSI) crime investigators. What is the evidence that a man had died? Let's examine the burial shroud which was found in the tomb. There are obvious signs of blood wherever there were wounds to the body. So to leave blood on the shroud, it must have come in contact with the body. Are there any witnesses to this man's death? Yes, there are four written accounts and they agree on what happened. The man was arrested early in the morning and after a brief hearing and trial, the man was found guilty by a mob and scourged brutally over his entire body and crowned with piercing thorns. He began to bleed profusely from his head and scourge wounds and blood covered his entire body. It was reported that he lived another six to eight hours dying on a wooden cross at 3:00 p.m. What would be the consequence to his body if it were covered in blood for that time period? According to medical microbiologists, his own skin bacteria would reach levels rarely if ever seen before. These bacteria produce sticky polysaccharides that function as glue. They would coat his body all over and if the body was washed briefly with water they would coat the body evenly. After the linen shroud was placed over the body, the sticky polysaccharides would seal the linen to the body particularly where bones could serve as anchor sites. In addition, the polysaccharides would seek any remaining water in the body further sealing it to the body. Since a body was not found in three days, CSI speculates that the body was removed from the linen and secured elsewhere, but the linen was left behind in the tomb. For what purposes can only be speculated. But left on the linen was the evidence of body contact. Examination of the linen revealed a massive presence of bacteria. The specific color of the linen was attributed to the oxidation of fatty acids and phospholipids leaving a permanent insoluble record of the crucifixion and the death of the Man of the Shroud.

Image Formation

Regarding the mechanism of image *formation*, the presence of an unusually large number of bacteria on the skin is responsible for image formation on a linen surface with the attributes of a photograph. As it concerns the Shroud, the basis for the unusual number of skin bacteria can be attributed to the prolonged period of bleeding from the head area as well as the rest of the body (probably six to eight hours), which allowed normal skin bacteria, particularly *Staphylococcus epidermidis* (*S. epidermidis*), to continue to grow at 37°C (body temperature) and reach extremely high levels. This is a natural phenomenon but it required specific events to occur in a sequence. Under these defined conditions, an image will always form.

Denial of these processes and their physical consequences is not a scientifically viable option: the presence of high levels of *S. epidermidis* on the body at the time of death is a certainty. **To say otherwise is to disregard the biology of the human body**. In the case of the crucifixion, as recounted in the Gospels, there is nothing that would have prevented the growth of *S. epidermidis* over the body of Jesus. It will happen every time under the described conditions. And given that the image on the Shroud is the image of a crucified man, this **is not another hypothesis for how the image formed on the Shroud of Turin.** It is the only natural explanation that takes into account the entire journey from Jesus' arrest in the morning through his crucifixion to his death and burial.

Shroud Preservation

With regard to the *preservation* of the Shroud, growth of innumerable **other** microorganisms over time stabilized the surface and interior regions of the linen fabric. They have, in effect, held it together making the Shroud a self-repairing artifact. Fungi, in particular, produce long web-like mycelial elements that would interweave through the Shroud fabric. It is not necessary to examine the Shroud linen to make this observation. All surface objects on the earth are coated with microorganisms, both living and dead. In fact, we now know that microorganisms are found everywhere on earth including deep within the earth's rocks and in superheated thermal vents deep within the oceans. We now know that the billions of microorganisms that constitute our microbiome are essential to our survival. Likewise, the microorganisms that attached themselves to the Shroud over the centuries kept it intact.

The bacteria that had contributed to image formation on the Shroud had long ago died as nutrients from the skin and body became exhausted. However, other bacteria with fewer *Staphylococcus epidermidis* nutritional demands grew on the skin bacteria utilizing some of their disintegrating macromolecules. In time, bacteria and fungi completely covered the entire Shroud and have colonized every available site on the linen. A few of these bacteria could have produced intracellular and extracellular polymers with the properties of a transparent plastic-like material. The most common plastic-like polymer produced by some bacteria is a poly-β-hydroxyalkanoate, which has the tensile strength of ordinary plastic and is widely used in Europe as biodegradable plastic.

Thus various bacterial products have served to protect the linen from decomposition. However, oxidative processes due to live microorganisms and directly from oxygen itself will continue and, over a period of time, it will become increasingly more difficult to distinguish between image and non-image areas. All microorganisms require several factors for growth including an energy source, specific nutrients, and moisture. In some cases, an energy source can also serve as a nutrient source. Atmospheric gases, such as nitrogen, carbon dioxide, carbon monoxide and others, can serve, in some cases, as energy sources and sole sources of nitrogen and carbon for certain bacteria which are referred to as autotrophs. Heterotrophs are bacteria that utilize preformed carbon such as that found in dead bacteria. Some bacteria are aerobic and require molecular oxygen for metabolism. Molecular oxygen is well known to directly oxidize organic molecules over a period of time and thus contribute to the yellowing appearance of the linen.

The diversity of microorganisms residing on and in the Shroud is reflective of the type of environmental niche provided by the Shroud linen. In any fabric there are numerous ecological niches suitable for a large array of different species. On the surface layer exposed to the atmospheric oxygen, strict aerobic bacteria would be expected along with microbes that could also grow in a reduced oxygen environment when oxygen levels fall at the surface as cells accumulate. Directly below these cells would be microbes that need much less oxygen. Deeper still and inside the lumen canals of the fibers would be those microbes that are actually inhibited by oxygen and grow only in its absence and are known as strict anaerobic microbes.

The initial colonizing microorganisms of the Shroud linen not associated with the image would likely be those able to use trace amounts of organic material adhering to the fabric. This material would come largely from humans handling the flax fiber as well as those involved in the weaving process. A recent investigation examining bacteria colonizing polyester and cotton clothes following a fitness session indicated that ninety-one isolates representing thirty-one unique species were obtained from aerobic and anaerobic platings. Both Gram-positive and Gram-negative bacteria were isolated. The Gram-positive species were the skin-resident staphylococci, *Microccosus* spp. and *Bacillus* spp. Gram-negative bacteria found were *Acinetobacter* spp. and *Enterobacteriaceae* spp.[4]

Once these bacteria are present on a cellulose-based cloth surface, they may go through several growth cycles before exhausting available nutrients then becoming dormant and eventually dying. However, they now serve as a nutrient deposit for other microorganisms falling on the surface and taking up residence. One can observe in nature how the initial colonizing species may be somewhat limited in function but, over a period of time, the ecological site becomes increasingly complex and interdependent. This would be certainly true with the Shroud of Turin. Bacteria and fungi able to consume the organic nutrients of other microbes, as we saw, are referred to as heterotrophs. There is another entire group of bacteria that can utilize carbon dioxide (and carbon monoxide) as their sole source of carbon and along with trace elements grow and reproduce. These, as noted, are autotrophic bacteria and are certainly represented in the Shroud microbiome. Because of the greatly increased energy requirements to supply all their carbon needs, they would grow much slower than heterotrophic bacteria and thus be at a competitive disadvantage. However, the age of the Shroud of Turin is extensive by any measurement and time becomes less of a limiting factor.

With regard to the survival of the Shroud in its current form into the future, because of all these biological and chemical processes, moisture and oxygen must be excluded from contact with the Shroud. Although some bacteria can grow in the absence of oxygen (something that critics of the embedded microbiome paradigm have failed to note), they still have a requirement for moisture, so the absence of both moisture and oxygen is necessary. The choices for exclusion of oxygen include either using a vacuum or an inert gas, such as argon. A number of techniques are available for excluding moisture. The continued exclusion of moisture over a long period of time will eventually result in the death of all vegetative microbial life forms with the exception of spores, which may exist for an undefined period of time, but causing no further damage. Exclusion of water will result in a loss of luster in both the image and non-image areas. This is due to removal of water from microbial polymers, such as polysaccharides, which have a high affinity for water, and contribute to the glistening or moist appearance. This removal of moisture should not be considered detrimental to the Shroud's integrity.

Dating, et al.

Once we recognize the Shroud to be a biological artifact, we can better appreciate the inescapable challenges in accurately dating it. In point of fact, the very nature of the mechanisms involved in creating and preserving the Shroud image makes it all but impossible to accurately date the underlying material. Dating the Shroud simply becomes a matter of dating the various microorganisms and their remains that have interacted with it across its history! These "contaminants" are physically speaking now a part of the Shroud because the material comprising bacterial and fungal cell walls has the same chemical structure as the cellulose that is the primary component of linen. Just as a scientist, no matter how well trained, cannot "separate" you from your lifetime of meals, it is all-but-impossible for investigators to distinguish the Shroud's original material from later squatters who are not only bio-chemically identical but now a part of it.

Thus, radiocarbon-type dating initiatives are doomed from the start because there is no way to distinguish between the beta-1,4-glucose of the original cellulose linen and the beta-1,4-glucose of the bacteria and fungi. This is because the cell walls of bacteria and fungi at their basic biochemical level are beta 1,4-glucose linked polymers.

There is another reason why it is well-nigh impossible to distinguish between the Shroud linen and its microbial population. One of the most important biological processes on earth, essential for the survival of living organisms, is the fixation of atmospheric dinitrogen into organic nitrogen thus enabling amino acids and proteins. Only certain bacteria have the ability to play a role in these processes. So finding fixed nitrogen on something like the Shroud of Turin would be like gold to bacteria falling onto the Shroud. Consequently, N-acetyl groups on peptidoglycan and chitin molecules as well as the amino acids in the peptidoglycan bridges would be a gold-mine for bacteria and fungi looking for fixed nitrogen. Now bacteria and fungi have enzymes that can remove these N-containing compounds and make them available as nutrients. The bottom line is that the cell walls of bacteria and fungi already on the Shroud would look exactly like the plant based cellulose that makes up the linen. With the identification of bacterial and fungal cell walls as really cellulose (after the enzymatic removal of the nitrogen components) we recognize the image area as being cellulose. The straw yellow color derives from the insoluble oxidized fatty acids from the bacterial, fungal, and skin cells.

As we will see in more detail, there are two unique aspects of the new paradigm of the Shroud's embedded microbiome that magnify its explanatory power: you can "see for yourself" that its claims are constantly verified in the physical world as well as "do it yourself" in the sense of testing out the truth of its assumptions *on* yourself. In fact, anyone who rejects the paradigm has to first address these two features of everyday experience if they want to be taken seriously. For precisely this reason, the burden of proof is very much on the skeptic.

In the present work, we have sought to go beyond the peer-review process to present experiments that can be replicated and verified by anyone. Although peer-review helps separate the wheat from the chaff, replicability is the gold standard in science. In recent years, a number of studies have reported that it is not sufficient for a paper to be peer-reviewed. One such report was published in *Nature*. Two eminent drug development scientists announced that when they sought to "confirm published findings" in "fifty-three papers [that] were deemed 'landmark' studies," they found that "scientific findings were confirmed in only 6 (11%) cases." They highlighted what they deemed to be the root of the problem. "What reasons underlie the publication of erroneous, selective or irreproducible data? The academic system and peer-review process tolerates and perhaps even inadvertently encourages such conduct."[5] Given this background, our focus here has been on experiments that can be replicated.

Some critics have said that a first century date for the Shroud would mean that it is 60% "contaminated" with microbes. This "criticism" proves nothing more than ignorance of the microbial world. As we will see for ourselves, these levels of contamination are not only not unusual but to be expected in organic artifacts exposed to common microorganisms.

For clarity's sake, it should be noted that the paradigm of the microbiome outlined here does not endorse the bioplastic coating theory of the pediatrician Leoncio Garza-Valdes or his claim that the blood on the Shroud contains "the DNA of God." Garza-Valdes claimed that a novel polyester-producing microorganism covered the Shroud with a bioplastic coating. Although such bacteria do exist and, as mentioned, are used to make the biodegradable "plastic" used in Europe, we take no position as to whether or not they are to be found on the Shroud. The skin bacteria from the Man of the Shroud are sufficient to explain the formation of the image.

1.5 The Embedded Microbiome that Created the First Photograph in History

"The shroud is a remarkable artefact, one of the few religious relics to have a justifiably mythical status. It is simply not known how the ghostly image of a serene, bearded man was made. It does not seem to have been painted, at least with any known historical pigments."

Philip Ball. Editor, *Nature* (17 years after *Nature's* Carbon 14 dating report)[1]

Scourging and implantation of the Crown of Thorns caused continuous bleeding.
↓
The skin bacterium *Staphylococcus epidermidis* (SE) began dividing every 30 minutes because of the nutrients in the blood and ideal body temperature.
↓
SE became the dominant bacterium because massive amounts of sweat (sodium chloride) produced during crucifixion inhibited the growth of other bacteria.
↓
SE reached millions of cells per square centimeter on the skin, levels not normally seen, during the 6-8 hour saga of suffering ending in crucifixion.
↓
After death, the body was washed with water spreading SE over the entire surface of the body.
↓

Once the linen was placed over the body, the sticky polysaccharides produced by SE pulled the linen tightly onto the surface of the body.

↓

The water-seeking linen then pulled the excess bacteria and sticky polysaccharides along with dead epithelial cells onto the linen surface.

↓

Upon drying and exposure of the linen to air (oxygen) after the inexplicable disappearance of the enclosed body, unsaturated fatty acids from the cell membranes of bacteria and dead epithelial cells became oxidized and turned yellow creating the image. Variations in amounts of unsaturated fatty acids created lighter and darker areas providing for the photographic negative and 3D attributes of the image (the yellow image is the photographic positive image).

We cannot understand the Shroud without a grasp of its embedded microbiome. There are two dimensions to this microbiome: the primordial microbial action that created the Shroud's image and the subsequent microbiota that preserved it and, in fact, became an integral part of it. Here we are concerned with the first, namely the formation of the image.

It is no exaggeration to say that the image created by the primordial microbial action, the Jesus Microbiome, is literally a photograph, in fact the first photograph in history.

There were four players in the creation of this "photograph":

- The body of the Man of the Shroud
- His skin bacteria (specifically *Staphylococcus epidermidis*)
- The linen in which the body was wrapped
- The exposure of the linen to oxygen

Three processes enabled the photograph:

- Creation of sticky polysaccharides by the skin bacteria
- Oxidization of unsaturated fatty acids which are in skin bacteria and skin cells

- Variation in concentration of these oxidized fatty acids leading to the features of the resulting image

Photography, invented in 1839, used light and chemistry to achieve its results. The photographs of that time, all the way to the invention of digital photography, were images made by the action of light on light-sensitive surfaces. The underlying action is the chemical imprinting of an image that accurately represents the thing or person being imaged.

The same chemical principle applies in the formation of the Shroud image. Here, the unusually large number of bacteria on the skin of the man wrapped in the Shroud is the primary driver in the formation of an image on its linen surface that has the attributes of a photograph. Whereas, in classical photography, images are produced by the effects of light on the oxidation of photosensitive chemicals, in the case of the Shroud the image is produced through the oxidation of unsaturated fatty acids.

As shown in a later chapter, this process of image creation from skin bacteria can be replicated here and now complete with nose, lips, eyebrows, nails, etc.

Background

First some background. It is now known that over a thousand species of bacteria reside on human skin. Among these are species of *Staphylococcus*, *Corynebacterium*, and *Propionibacterium*. *Staphylococcus epidermidis* (*S. epidermidis*) is the most common bacterium on human skin. Nutrients for these bacteria are largely obtained from the disintegrating epithelial cells on the surface of the skin and moisture provided by sweat glands. The nutrients and moisture levels are generally present in limiting amounts preventing rapid growth of skin bacteria. The bacteria that are normally found on human skin completely coat the surface and are estimated to range from hundreds to thousands per square centimeter. An important function of these normal flora bacteria is to occupy all sites on the skin thereby preventing more pathogenic microorganisms from gaining a foothold.

One of the properties of *S. epidermidis* that allows it to efficiently coat the skin surface is the ability to produce glue-like polysaccharides, which provide for its strong attachment to receptors on epithelial cells. Our modern hygienic bathing procedures result in the daily removal of accumulated dead epithelial cells along with their attached bacteria without leaving visible signs of their presence on towels or linen surfaces. However, when the skin surface becomes covered with rich nutrients from blood, serum, or a culture medium, *S. epidermidis* present on the skin can grow to extremely high levels. Bacteria present at 1,000 colony forming units/cm^2 on the skin and dividing every 30 minutes can grow to 20 million units/cm^2 over an 8-hour time period.

Sequence

To return to the image formation on the Shroud, the skin of the crucified Man of the Shroud would present a means for transferring readily detached epithelial cells in a biofilm containing *S. epidermidis* to the linen surface. The sequence in the transfer of the skin cells and bacteria to the linen is dictated by very specific environmental conditions. The crucifixion and burial of Jesus was a unique event in the history of humanity and the exact environmental conditions at that time can never be duplicated. Thus, an exact replication of the Shroud of Turin is impossible. However, the process of transfer of unusually high levels of skin bacteria from skin surfaces to linen resulting in image formation can certainly be examined.

We noted earlier that the extensive bleeding of the Man of the Shroud would have resulted in an extraordinarily high number of skin bacteria and that these bacteria would produce sticky polysaccharides functioning as glue. When linen is wrapped around the body, it becomes sealed to the body by the sticky polysaccharides with the bones serving as anchors. The massive numbers of bacteria along with skin cells on the body have a greater attraction for the linen because of the affinity of the bacterial polysaccharides for the moisture in the linen. In addition, both the linen and the polysaccharides are drawing residual water from the body. This combination of bacteria, skin cells, and polysaccharides is the photographic emulsion. When we expose this emulsion to oxygen, the unsaturated fatty acids in bacterial and skin cell membranes become oxidized and turn yellow.

Chemically speaking, oxygen oxidizes unsaturated fatty acids by adding an atom of oxygen that breaks double bonds. This process produces an oxidation product that has a straw yellow discoloration

The image itself is "formed" by higher and lower concentrations of oxidized unsaturated fatty acids (bacteria and skin cells). If you look at the positive image, the yellow areas of the face are higher concentrations, the darker areas are even higher concentrations, the lighter areas are lower concentrations. Of course, this is reversed in the negative image. So the varying concentrations of bacteria and skin cells (unsaturated fatty acids) provide for both the image and its 3D structure.

This is the same in classical photography: when we expose the photographic emulsion containing silver halides to photons of light, a chemical reaction occurs resulting in an image being deposited on film or glass plate. So oxygen and light have the same function – to cause a chemical reaction resulting in the development of images.

You might imagine that there is a struggle for where the bacterial and dead skin cells are going to be located. Will they stay with the body or be drawn onto the linen? The linen wins out because it has a stronger affinity for water. The uniqueness of the process required bacteria that would produce a sticky polysaccharide that confined the image to only the subject "photographed." Not all bacteria produce sticky polysaccharides. The bacteria must be able to grow to unusually high levels to produce enough unsaturated fatty acids to be visible when oxidized. This can only occur when sufficient nutrients are provided over a long period of time. This is usually not seen naturally. Thus, the Shroud of Turin image is unique and no other examples have occurred thus far in recorded history.

Due to the initially moist linen or subsequent exposure of the linen to moisture, diffusion of water-soluble nutrients (blood, serum and culture medium components) would allow skin bacteria to grow in areas not in direct contact with the skin surface. However, the extent of diffusion of organic materials would provide a definite boundary to the image. Skin bacteria, such as *S. epidermidis*, require complex nutrients and therefore are not able to grow on linen without nutrients.

The most readily oxidizable organic materials that contribute to a yellowing appearance are lipids (fatty acids) and some pigments that are susceptible to oxidation by molecular oxygen. However, other bacterial components, such as the cell wall peptidoglycan, are not as readily oxidized but will contribute to color development over a period of time. Bacterial cell wall peptidoglycan is the main structural component of the cell wall of Gram-positive bacteria such as *S.epidermidis* and comprises as much as 50% or more of the cell by weight. Bacterial peptidoglycan has the same basic structure as that of cellulose, namely, β-1,4-linked glucose polymers. It is only distinguished by the presence of N-acetyl groups, phosphoenol pyruvate groups and several amino acids. Other cell material would include DNA, RNA, and protein as well as extracellular polymers, such as polysaccharides, and other types of organic material. Over long periods of time, periodic increases in moisture content would allow microbial enzymes to break down the larger polymers into smaller molecular weight material, which would be more prone to oxidation, and thereby increase the intensity of the image.

The bacteria are also responsible for the three-dimensional nature of the Shroud image discovered by the STURP team. Because of the presence of the bacteria, more is involved than a simple wrapping of the linen around the body. Both the body and the linen participate in parallel processes. The body is coated with the bacteria seeking remaining traces of water within the body. The wrapping of the body with a water-seeking object, namely the linen, pulls the linen very tightly onto the moist body. The anchor points are the bony framework of the body. As drying occurs the linen exerts a greater force for excess skin bacteria and their polysaccharides pulling them from the body. Thus, it is not a simple wrapping but a dynamic interaction between the body and the linen both competing for available water. And, as a result, the cloth encodes the information of its distance from the various parts of the body it wrapped.

Emulsion

At a more technical level we can compare classical photographic emulsions with the "photographic emulsion" found on the Shroud.

The term photographic emulsion refers to a colloidal mixture usually containing gelatin as a base coating on a surface. Within the colloidal mixture are evenly distributed insoluble silver halides, which are extremely light sensitive. When photons of light enter the open camera lens, they strike these light-sensitive compounds and instantaneously cause chemical reactions which ultimately result in the development of an image observed by the camera lens. In the case of the image found on the Shroud of Turin, the colloidal mixture would be the unusually thick coating of human skin bacterial cells containing oxygen-sensitive and highly reactive double bonds of the evenly dispersed fatty acids of the cell membranes. Upon removal of the linen cloth from the body (opening of the camera lens) the previously excluded oxygen molecules instantaneously react with unsaturated fatty acids causing chemical reactions leading to oxidation of the fatty acids and permanent yellowing of the linen material. As oxidation proceeds over time the initial yellow coloring on the linen fabric may deepen in intensity and develop a more straw-like color. One can also observe an oxidative-process in old photographs over time as they develop a yellowing of the original black and white photograph.

The image found on the Shroud of Turin is the first recorded naturally occurring "photograph" in human history. The intensity of the image is directly correlated with the amount of bacteria found on the skin of the deceased. The amount of bacteria, in turn, is directly correlated with the amount of blood coating the body of the crucified victim. Based on the growth rate (division every 30 minutes) of the skin bacterium (*Staphylococcus epidermidis*), a time period of 6-8 hours as documented in the four gospels would result in tens of millions of bacteria on the skin surface at the time of death, more than sufficient to produce this remarkable "photographic" image.

It is, as we have said, an Instagram from the first century!

1.6 Paradigm Shift a.k.a. Game-Changer

Paradigm = model; template; "a theory or a group of ideas about how something should be done, made, or thought about" (Merriam-Webster)

Paradigm Shift = game-changer; "a time when the usual and accepted way of doing or thinking about something changes completely" (Cambridge Dictionary); "a fundamental change in approach or underlying assumptions" (Oxford Dictionary)

Game-changer = "something that completely changes the way something is done, thought-about, or made" (Macmillan Dictionary)

The paradigm proposed here is simple: like every organic artifact, the Shroud of Turin has its own embedded microbiome and the nature of this microbiome is evident from both the image on the Shroud and its preservation over centuries. The new paradigm is a shift from models derived entirely from physics or based wholly on speculation. As a result of it, we now KNOW purely from observational 21st century science, as opposed to the inferential speculation characteristic of dating methodologies like Carbon-14, **how** the image on the Shroud came to be, **why** "dating" it with conventional technologies is impossible and **that** the Man on the Shroud is the Man of Sorrows (Isaiah 53 and the Passion narratives of the New Testament Gospels). It is no exaggeration to say that modern observational science has SHOWN the Shroud to be what it is and how it came to be – without any recourse to supernatural explanations.

Seeing is believing!

But we cannot reach this new level of insight and understanding – we cannot "see" – without a paradigm-shift, a mental make-over. "Paradigm-shift" became part of the popular lexicon after Thomas Kuhn used it in *The Structure of Scientific Revolutions* to describe revolutions in science. "Normal science" involves scientific work done within a particular framework of assumptions, concepts and practices. A scientific revolution displaces the existing "normal science" framework of fundamental concepts, assumptions and practices and presents a new way of viewing and studying the same data. Copernicus "overturned" the idea that the earth is at the center of the Universe: the Copernician Revolution ended the reign of Ptolemy in astronomy. Likewise, quantum physics replaced Newtonian physics as the grid for describing the behavior of subatomic particles and fields. This kind of displacement of fundamental frameworks is what Kuhn called a paradigm-shift.

It is precisely such a radical change in the rules of the game that has transformed the scientific investigation of the Shroud of Turin. And it is a paradigm-shift occasioned by the introduction of microbiology, a new and particularly relevant branch of science, in studying the very same data. The Shroud microbiome unveiled by microbiology has turned out to be a game-changer. Its relevance to Shroud studies is the same as that of quantum physics to analysis of the microverse.

The relevance of microbiology becomes self-evident once we open our "eyes" (via a microscope!) to the microbial world. Microorganisms are ubiquitous and essential to nature and life. There are ten million microorganisms in a milliliter of ocean water, 40 million in a gram of soil and tens of trillions in or on our bodies. Organic objects are in constant interaction with these organisms. Moreover, wounds leave us vulnerable to the invasion of numerous other kinds of bacteria.

We know that the Shroud of Turin was a shroud and that it was traditionally believed to have been used to wrap the corpse of a severely wounded man. If this man was the Jesus of the Gospels, we know that he was a human being with normal skin bacteria. If his body was wrapped in a shroud, as the accounts of his death report, there can be little doubt that his microbiome interacted with his burial cloth.

In this context, it is puzzling that microbiologists were never called to participate in the 1988 dating study of the Shroud that sought to evaluate the claim that it was Jesus' Shroud. It was only in the Nineties, fairly late in the game, that one of us (SJM) was asked to evaluate the Shroud from the standpoint of his discipline, microbiology. But once introduced, microbiology has ignited a paradigm shift in Shroud studies – one which has still to be comprehended by both friends and foes of Shroud studies.

The impact of the microbial world on the Shroud is obvious at the three levels we have reviewed and will expand on:

- Its role in the formation of the image on the Shroud
- Its function (amazingly) in "preserving" the Shroud
- Its relevance for any effort to "date" the Shroud

Inevitably there will be questions and critiques. For instance, those concerned with the dating question might ask: How much contamination would be required to skew the dating process and why isn't it visible? And wouldn't such contamination have been eliminated in the cleaning carried out before the Carbon 14 tests? We will get to these and other questions but no answer will satisfy those who refuse to move beyond the old paradigms. Such obstinacy is nothing new in the history of science as Kuhn has documented: often the practitioners of "normal" science were unwilling to enter the thought-world of the revolutions that left them behind. So it has been in the case of the new paradigm in Shroud studies.

Remarkably, but perhaps predictably, the sharpest critics of this new paradigm have been *defenders* of the Shroud's antiquity and authenticity! Paradoxically, at the same time, Harry Gove, the co-inventor of the AMS (accelerator mass spectrometry) radiocarbon dating methodology, was actively engaged in exploring the microbial contamination of the Shroud with a co-author of this book (SJM). This is not to say that the polemical Shroud skeptics have shown any inclination to consider the new paradigm. But the fire and fury from both sides of the Shroud debate indicates that the newness of the model is the issue and not necessarily its results. If you approach everything from the standpoint of physics, for instance, you will miss facts and insights that fall outside its purview. Amusingly criticisms from both sides of the aisle proceed along the same lines and have the same fundamental flaw: a failure to grasp the all-pervasive change of perspective demanded by the paradigm of the embedded microbiome.

It should be said here that some bodies of data remain the same across different paradigms. Gravity was experienced by both Newton and Einstein. It was their explanation for the phenomenon that was different. In the case of quantum physics, the paradigm-shift was occasioned by new data. The paradigm of the embedded microbiome addresses both previously known data (Shroud 2.0) and the new universe of microbiota and its role in the formation and preservation of the Shroud image.

1.7 It Seemed Like a Good Idea at the Time – Searching for the Carbon "Fingerprint"

It is appropriate now to revisit the issue of dating raised in 1.2. Radiocarbon dating seemed the obvious choice in determining the age of the Shroud. The Carbon 14 (C14) dating methodology, derived from the measurement of the radioactive carbon in an organic substance, is one of the most remarkable achievements of modern times – a breath-taking testament to human ingenuity.

As is only too well-known, the C14 dating of the Shroud, carried out in 1988, came to a negative conclusion. The *Nature* paper on the study was emphatic in its conclusion: "The results of radiocarbon measurements at Arizona, Oxford and Zurich yield a calibrated calendar age range with at least 95% confidence for the linen of the Shroud of Turin of AD 1260 - 1390 (rounded down/up to nearest 10 yr). These results therefore provide conclusive evidence that the linen of the Shroud of Turin is mediaeval."[1]

The "95% confidence" level is puzzling since "rehearsals" before the final test had resulted in outliers where one sample was dated a thousand years too young and another a thousand years too old.[2] So it would behoove us to delve deeper into the radiocarbon investigation rationale and its relevance to the dating of the Shroud.

Briefly, the rationale derives from the laws of physics. Carbon is normally found in one of three isotopes: C12, C13 and C14. C14 (Carbon 14) or radiocarbon is a radioactive isotope of carbon that forms naturally through the action of cosmic rays on the atmosphere. But it has also been artificially generated principally via the nuclear tests of the Fifties and Sixties. Radiocarbon atoms act like other carbon atoms with the difference that they weigh more and decay into nitrogen over thousands of years. The ratio of the various carbon isotopes remains quite stable with C12 comprising 99% of all the carbon on earth and C14 .0000000001%. The half-life of C14 (the period in which half of a given sample decays) is 5,730 years. These two features (ratio and half-life) make C14 a useful tool in measuring the age of radiocarbon "containers" (us, for instance!).

C14 becomes a part of living things principally through photosynthesis when radioactive carbon dioxide (formed as atmospheric radiocarbon interacts with oxygen) is processed by plants and these in turn are eaten by animals. C14 also becomes part of living organisms through the pyruvate carboxylase enzyme. When plants and animals die, the ingestion of C14 comes to a stop and the radiocarbon they "contain" at that point begins to decay. Since the C14 half-life is known (5,730), measurement of the radiocarbon in a "deceased" party tells you when it expired and thus how long it has been in existence. The less it has of C14, the older it is. It is 5,730 years old if it has only half its original C14. Every subsequent 5,730 years, the C14 present is reduced by half.

The older "beta counting" technique of determining a sample's C14 by measuring its radioactivity was replaced in the Seventies by the AMS (accelerator mass spectrometry) methodology which measures the actual number of C12 and C14 atoms in it. AMS arrives at its age projection by analyzing the ratio of the carbon isotopes measured.

Accuracy Limitations

But the accuracy of the dating can be affected by such factors as variations in the atmospheric C12/C14 ratios as happened after the 20th century nuclear tests. Of more concern, however, is contamination of the sample being measured. If any modern carbon is added to a sample, it will appear younger than it is. This makes decontamination a high priority in any radiocarbon test.

Demonstrably wrong carbon-dates are not uncommon as archaeologists and others have testified. A 1990 paper in *Nature* drew attention precisely to the measurement skewing challenge: "Since 1947, scientists have reckoned the ages of many old objects by measuring the amounts of radioactive carbon they contain. New research shows, however, that some estimates based on carbon may have erred by thousands of years. ... Scientists at the Lamont-Doherty Geological Laboratory of Columbia University at Palisades, N.Y., reported today in the British journal Nature that some estimates of age based on carbon analyses were wrong by as much as 3,500 years. ... *[S]cientists have long recognized that carbon dating is subject to error because of a variety of factors, including contamination by outside sources of carbon.*"[3]

University of Zurich researcher Willy Woelfli, a member of the Shroud carbon dating team, has himself drawn attention to the limitations of radiocarbon dating: "The existence of significant indeterminant errors can never be excluded from any age determination. No method is immune from giving grossly incorrect datings when there are non-apparent problems with the samples ... the results illustrated [in his presentation] show that this situation occurs frequently."[4]

In another paper, he highlighted contamination and other sources of error: "The major problems in radiocarbon dating with AMS are very similar to those observed in conventional radiometry: The fact that ^{14}C abundances in a sample can be evaluated reproducibly with high accuracy is no guarantee that the corresponding radiocarbon date is per se correct. Natural and/or artificial sample contaminations, such as indicated above, sample inhomogeneities, or uncertain origin of material are well known and frequent sources of error, particularly in small sample dating."[5]

The 1988 Test

Which brings us to the 1988 Shroud-dating study. Even at a superficial level, any attempt to date the Shroud posed manifold problems because of its public exposure since the Middle Ages. These were compounded by the controversy over the changes in protocol for the actual test. The protocol for the carbon dating was established by an inter-disciplinary committee in order to avoid the rehearsal errors. It

was agreed that the study's accuracy depended on certain key parameters being put in place:
- The actual carbon-dating should include two measurement methodologies: AMS (described earlier) and the Small Proportional Counter (SPC) system: the results generated by each would serve as a check on the other.
- The samples would be tested at seven different labs.

But by the time of the actual test, the agreed-upon protocol was summarily abandoned:

- The SPC was eliminated and only the AMS methodology was deployed.
- Only three labs would be included in the study.

Two members of the committee had asked that the samples dated be taken from different sections of the cloth but this request was overruled. Samples were taken from only one section of the cloth, a section likely to have been the most contaminated since it was known to have been constantly handled by those who displayed the Shroud from the Middle Ages.

Moreover, the study excluded all those with any knowledge of the Shroud, in particular the STURP team. Other non-Shroud radiocarbon studies included not just the labs but the archaeologists, biologists, textile experts and the like who had specialized knowledge of the samples being studied. These experts acted as cross-checking resources against false readings. But the Shroud dating team did not include anyone with specialized knowledge of the cloth under study.

Contamination

Even more troublesome, there was no way to overcome the most formidable obstacle to the radiocarbon dating enterprise: contamination. If a sample is not effectively decontaminated, any measurement of its age is by that very fact worthless with respect to accuracy. If a three thousand year old sample includes contaminants from one thousand years ago, radiocarbon dating will show the sample to be younger than it actually is. In the case of the Shroud investigation, a single solvent was used by all three labs as a decontaminant without sufficient consideration of its effectiveness against the Shroud's known contaminants let alone the microbiological dimension.

Historian Dan Scavone recounts the Shroud's known contaminants: "It is known that Shroud contamination includes oil, wax, tears, incense, and the smoke from a fire in 1532, when an abundance of carbon of that date must have thoroughly saturated the Shroud. One need not accept the suggestion that the Shroud's carbon nuclei would be altered by such an explosion of steam, but this event in the Shroud's 'thermal history' renders it unique in all the history of C-14 dating. The Shroud is thus an extremely unusual instance in which much later substances have been in contact with the sample at elevated temperatures; in this and in being handled for 600 years it is immensely different from objects retrieved from the ground of an archaeological dig which have been untouched for centuries."[6]

The Shroud has a history of surviving various fires, the worst of which was in 1532. The significance of the 1532 fire with regard to contamination has, perhaps, not been sufficiently appreciated. The "unique ... thermal history" of the Shroud was highlighted by textile expert John Tyrer, a professor at the University of Manchester: "The heat inside the silver reliquary must have been intense probably reaching a temperature of 900 degrees C., the temperature of molten silver. In these circumstances, moisture in the Shroud would turn to steam at superheat, trapped in the folds and layers of the Shroud. Any contaminants on the cloth would be dissolved by steam and forced, not only into the weave and yarn structures, but also into the lumen and molecular structure of the fibers of the flax ... Contaminants would have become part of the chemistry of the flax fibers themselves and would be impossible to remove satisfactorily by surface actants and ultra-sonic cleansing."[7]

The sample for the study was itself problematic. In a paper in the American Chemical Society's journal, chemist Alan Adler observed that "only a single sample was taken in the lower corner of the main cloth of the frontal image below the so-called sidestrip from the selvage edge in an obviously waterstained area just a few inches from a burn mark. The selvage edge was trimmed off before portions of the sample were divided among participating laboratories. Whether such an obviously contaminated sample is truly representative of the rest of the cloth is clearly questionable and the accuracy of the reported date is certainly doubtful."[8]

Ian Wilson and Barry Schwortz report that "It is a matter of firm record that the sliver of Shroud that was taken for the 1988 radiocarbon dating was snipped from its top left-hand corner, one of the two corners by which it was traditionally held up for exposition over the centuries. ... In countless engravings of Shroud expositions back through history, bishop after bishop can be seen clutching the Shroud at this very point. Now, as microbiologists are fond of demonstrating, microbes from even the cleanest hand will grow impressive colonies in an agar dish in a matter of days. So, if there is any point on the Shroud on which the maximum amount of microbiological contamination could be expected to have accumulated, it would have to have been these corners."[9]

Testing C14 Testing

Particularly relevant here is the major review of the accuracy of carbon-dating carried out a year after the Shroud investigation. Cambridge art historian Thomas de Wesselow reports that "In 1989 Britain's Science and Engineering Research Council (SERC) decided to conduct a trial in which the carbon-dating technique itself would be tested. Thirty-eight laboratories were involved in the trial, each being asked to date artifacts whose age was already known. ... The findings, reported in *New Scientist*, under the headline 'Unexpected errors affect dating techniques," were salutary. It was found that 'the margin of error with radiocarbon dating ... may be two to three times as great as practitioners of the technique have claimed ... Of the thirty-eight [laboratories], only seven produced results that the organizers of the trial considered satisfactory.' In other words, about 80 per cent of the labs failed the test." The AMS methodology used for the Shroud test "'came out of the survey badly ... some of the accelerator laboratories were way out when dating samples as little as 200 years old.'"[10]

The head of the STURP team illustrates these dangers with a list of well known carbon-dating discrepancies[11]:

> Carbon from any other source is considered to be "contamination". Such "contamination", if it is enriched with C-14 above the original atmospheric CO_2 level, will always lead to anomalous results if it is not removed before testing. Examples of famous cases of possible "contamination" not removed by the standardized method of cleaning used by radiocarbon laboratories are many, including the following:
>
> - Dating of living snail shells to be twenty-six thousand years old. [4]
> - Dating of a newly killed seal to be thirteen hundred years old.
> - Dating of one-year old leaves as four hundred years old. [4]
> - Dating a medieval Viking horn to the year 2006. [4]
> - Dating wrappings of an Egyptian mummy a thousand years younger than the body they wrapped.

De Wesselow observes that there is "a vast discrepancy, then, between the popular perception of carbon dating as infallible and its true scientific status. The fact is that carbon-dating results are often wrong, that the claims made on behalf of carbon-dating are often inflated, and that the AMS technique used in 1988 to date the Shroud is (or was) particularly error prone. The purveyors of any technology, carbon dating included, are inclined to exaggerate its power and usefulness. Also, being physicists, so not embroiled in the business of making historical sense of their findings, they probably have a tendency to underestimate the method's rate of failure. Those responsible for the historical interpretation of ancient artifacts, usually archaeologists, are the ones who decide whether or not to reject carbon-dating results. But, because archaeologists were excluded from the 1988 testing of the Shroud, scientific caution was thrown to the winds when the results of the high-profile test were announced."[12]

New Light on the Flawed Foundations of the 1988 Study

In 2019, the Oxford University journal *Archaeometry* published an explosive new analysis of never-before revealed documents underlying the original report on the 1988 Carbon 14 study. The analysis concluded that the study was fatally flawed with respect to both its methodology and execution thereby nullifying its inference that the Shroud originated in the Middle Ages.

The raw data behind the Carbon 14 study had never been released to the public despite repeated requests. Then, in 2017, Tristan Casabianca, a French researcher made a Freedom of Information Act request to the British Government. In response, he received hundreds of pages of documentation. After two years analyzing the documents, Tristan and his Franco-Italian research team published their negative evaluation "The tested samples," said Casabianca, "are obviously heterogeneous from many different dates. There is no guarantee that all these samples, taken from one end of the shroud, are representative of the whole fabric. It is, therefore, impossible to conclude that the Shroud of Turin dates from the Middle Ages."[13]

Casabiance noted that their report appeared in a journal published by the same "Oxford University Department, which dated [the Shroud in] 1988. Some commentators have immediately pointed out the irony that this represents, it shows above all that the challenge of dating is recorded at the highest level."[14]

Especially significant is the assessment of team member and co-author Emanuela Marinelli:

> The statistical analysis of the raw data, performed by the working group coordinated by Prof. Benedetto Torrisi, Professor of Statistics at the University of Catania, with Dr. Giuseppe Pernagallo, Dr. Tristan Casabianca and the

undersigned, published in *Archaeometry*, ... unequivocally confirms the inhomogeneity of the 14 C counts used for dating, probably due to a contaminant not removed by preliminary cleaning operations, a difficult problem to solve in the radio-dating of tissues, now well known and not considered quite important in 1988, as confirmed also by Prof. Paolo Di Lazzaro, physicist at ENEA in Frascati. The analyzed sample, chosen from a single point that was very polluted and was mended due to its peculiar characteristics, did not represent the entire sheet. Torrisi and Pernagallo emphasized that the strong inhomogeneities between the three laboratories and within the laboratories are alarm bells that confirm the non-statistical representativeness of the fabric fragments used in the sampling.

Already in 2012, the statistical tests conducted by Prof. Marco Riani, a statistician at the University of Parma, on the data published by *Nature* revealed that the datings provided by the three different laboratories were significantly different.

The statistical tests confirm not only that already on the official data the doubts on the agreeability were more than legitimate, but reinforces this thesis, bringing strong evidence of non-homogeneity as regards the raw data as well as for the datings provided by the Arizona laboratory alone.

The conclusions were summarized as follows by Prof. Torrisi:

We have no more doubts, the strong heterogeneity of the data leads us to affirm that the dating expressed on *Nature is not the correct one*.[15]

In her comments, Marinelli also referred to the importance of the microbiological dimension:

Many scholars were against submitting the Shroud to the dating with the 14 C method, due to the peculiarity of the find, which has gone through a thousand vicissitudes and is contaminated by many substances. Mold, fungal hyphae, candle smoke, sweat, fire, water, contact with more recent fabrics, restorations, may have significantly altered the linen, compromising the validity of the radiocarbon examination. Furthermore, the angle from which the sample was taken was one of the most manipulated parts during the exhibitions.

The silver reliquary containing the Shroud was enveloped in flames in the fire of December 4, 1532 in Chambéry; the high temperature in a closed environment can provoke exchange of isotopes that lead to an enrichment of radioactive carbon, making the fabric proportionally "younger". The reaction is favored by the presence of silver.

Some bacteria operating on the surface of the linen can, through their enzymatic activity, chemically bind alkyl groups to cellulose. These groups contain carbon derived from the local environment. Even if bacteria are removed by cleaning, cellulose changes remain. It should be emphasized that the transformation of linen due to fire and microbial action is chemical and not physical: therefore the solvents and cleaning techniques used by the radiological laboratories, which remove physical contamination, such as dirt, they do not remove the carbon-containing groups that have been added, because these groups form chemical bonds directly with the molecules of the cellulose itself.[16]

Margins of Error

Speaking as an art specialist, de Wesselow holds that "dating the Shroud to the Middle Ages makes it literally incomprehensible. …The Shroud is inconceivable as a medieval work of art and can be understood neither as a deliberate 'recreation' of Christ's burial cloth nor as a bizarre accident. The onus is on those who uphold the carbon-dating result to integrate it into a full and adequate description of the Shroud's origin – just as archaeologists would do with any other carbon-dating result. This they have been conspicuously unable to do."

De Wesselow says there is "a raft of reasons" to "reject the 1988 result": "it conflicts with all the evidence that points to the Shroud having been in existence long before 1260: the fact that the lignin in the fibers of the cloth has lost its vanillin, indicating that it is over 1,300 years old; the fact that the image derives from an actual victim of crucifixion, a practice outlawed in Christendom in the fourth century; the fact that the scourge marks testify to the use of a Roman flagrum and the fact that the technical features of the weaving and stitching conform to practices known in antiquity and not the Middle Ages. And this is only the evidence we have adduced so far."[17]

Errors in the dating results could have resulted from a number of factors. "In the vast majority of cases," writes de Wesselow, "when carbon-dating tests yield suspect results, it is because some natural process has interfered with the regular ticking of the radiocarbon clock. The most obvious explanation for the dubious carbon-dating result is that some form of contamination was present or that the level of C-14 in the material was otherwise enhanced ... The measurement errors caused by such processes can be spectacular – in the range of thousands of years.

"There is evidence that tests on linen are particularly prone to distortion. In the late 1970s Dr. Rosalie David of the Manchester Museum had samples from an Egyptian mummy carbon-dated at the British Museum, only to find that the bandages were dated 800 to 1,000 years younger than the body. She didn't believe the mummy could have been re-wrapped. ... In 1997 David co-authored an article ... in which new experiments conducted on ancient Egyptian ibis mummies were reported. It was found that 'there was a very significant discrepancy, an average of 550 years, between the dating of the mummy's linen wrappings and the mummy itself.' Two reasons have been suggested for the anomalous linen results: either that the porosity of the fibers makes them particularly susceptible to contamination, or that, because crop plants have a short life span, they reflect short-term fluctuations in the C-14 levels.

"The samples used in the 1988 results were cleaned using standard methods but, as [Harry] Gove remarks, 'One of the problems with small samples is that one never knows when the cleaning procedure was sufficient.' And he also makes the following point: 'All of the labs used the same cleaning technique, and if there's some kind of contaminant that was not taken care of, it would give the same answer to all three labs, and all three would be wrong.'"[18]

De Wesselow, himself an agnostic, concludes by asking: "Can we legitimately reject the carbon-dating result without determining exactly what went wrong? Of course we can. Archaeologists routinely dismiss 'rogue' radiocarbon dates out of hand. The success of a carbon-dating result should never be declared unilaterally; it is always measured against other evidence. The 1988 test may therefore be declared null and void, even though, without further study of the Shroud, it is unlikely we will ever be able to say definitively what went wrong.

"It is not just sindonologists who consider the carbon dating of the Shroud questionable. The thoroughly inconclusive nature of the 'conclusive evidence' trumpeted in the 1989 Nature article is acknowledged by the current head of the Oxford Radiocarbon Accelerator Unit (ORAU), Professor Christopher Ramsey: 'Anything is always provisional ... most scientific experiements are only verified by being repeated many times ... With the Shroud you're in a slightly difficult position, because obviously you can't go on dating it lots and lots of times. As a scientist, I'm much more interested in getting the right answer than in sticking to an answer which we came to before. ... There is a lot of other evidence that suggests to many that the Shroud is older than the radiocarbon dates allow and so further research is certainly needed.'"[19]

Explaining the Error

As referenced above, various other hypotheses have been proposed to explain the medieval C14 dating:

- the 1532 fire caused carbon enrichment
- the sample used in the study is not from the original Shroud but from a portion of it that was "repaired" in the Middle Ages

- a novel form of radiation resulting from the resurrection of Christ altered the physico-chemical structure of the Shroud.

The carbon enrichment hypothesis has yet to be verified by replication. The "repair" view has been defended by photographic comparisons between the test sample and the rest of the Shroud along with analysis of fibers reportedly from the sample area. But because of the fragmentary nature of the evidence adduced, this kind of argument has failed to convince even many Shroud supporters. The resurrection radiation claim, of course, falls outside scientific discourse given that it is thought of as a one-time event that cannot be replicated.

The objective in this book is not to question the value or applicability of radiocarbon dating. Despite its limitations, it is a remarkable scientific achievement that has made major contributions to the understanding of the past. Its inventors and practitioners must be applauded for their myriad breakthroughs. The primary concern here, however, is that in the case of the Shroud, radiocarbon dating faces formidable obstacles that have yet to be fully recognized. The very nature of the Shroud as a shroud that held a corpse makes it a unique kind of artifact that cannot be "dated" without decontamination techniques that were never applied in 1988. While other dating enterprises require archaeologists, the investigation of an artifact like the Shroud is especially dependent on the resources and insights of microbiology. And because this was not recognized at the time of the 1988 test, its results were inescapably flawed.

In brief, microbiology reveals the expected "behavior" of shrouds in contact with corpses and particularly of the kind of corpse that appears to have been wrapped in the Shroud. At the very minimum, decontamination would require the removal of the associated microbial populations and their "descendants." Since the required decontamination was not carried out – indeed could not have been carried out – in the C14 testing, its "dates" are inapplicable. Yet microbiology has a lot more to tell us than just the nature of the Shroud's contaminants. This is what we will explore in the next few chapters.

But this is the moral of the C14 story: Microbes are the elephant in the room when it comes to any ancient artifact. If you close your eyes or happen to be blind, the elephant will remain undetected. You have to regain your sight to be able to see it. As applied in this instance, you cannot detect a microbe if you do not have a device designed to detect microbes. As for the "cleaning" procedures adopted in the C14 study, if you have ticks in your house and you spray its every nook and cranny with pesticides that kill only larger insects, the ticks will continue to thrive and reproduce. In effect, you are letting the sleeping bugs lie.

The embedded microbiome paradigm, on the other hand, tells us what to expect.

1.8 See for Yourself – Launching a Microbiome Here and Now

Like any scientific theory, the paradigm of the embedded microbiome is demonstrable and replicable. Science is all about methodology that can be universally verified. There are two unique dimensions of the new paradigm that magnify its explanatory power. You can:

- "see for yourself" that its claims are constantly verified in the physical world
- "do it yourself" by testing out the truth of its assumptions concerning image formation *on* yourself.

Those who reject the paradigm have to first address these two *undeniable* features of everyday experience if they want to be taken seriously. For precisely this reason, the burden of proof is very much on the skeptic.

So how do you see it for yourself? At the most general level, all you have to do is take a shower, wipe yourself down with a white towel and then throw the towel into a closet. If you examine the towel a few months later, you notice discoloration on the towel caused by your skin bacteria and skin cells (this is much like the ring around the collar that runs across the neck of every shirt). Come back to it a year later and the microbial population on the towel has exploded. Ten years later the microbes are now an inextricable part of the towel. Recent studies have shown that in hostile conditions microbes can even cease metabolic activities and remain dormant for hundreds of years! And you won't know they are present without applying tools designed to track them down.

From these generalities, we move to the specific case of the Shroud. Here now is the replicable experiment performed by co-author Stephen J. Mattingly (SJM) demonstrating the kind and degree of microbial contamination undergone by the Shroud.

Microbial Contamination Case Study

A 100% cotton cloth fragment was cut into sections that weighed 250 milligrams (see Fig A). These fragments were washed with water several times and dried at 35 C. Re-weighing revealed no change in weight. *Bacillus subtilis,* a ubiquitous Gram-positive, spore-forming air contaminant, was grown overnight at 35°C on nutrient agar plates. Colonies of *B. subtilis* were mixed with 1-2 ml of Thioglycollate broth and applied to the clean cloths described above and incubated at 35°C overnight. The next day, the bacteria coated cloths were examined and weighed. The cloths became quite stiff and had a slight yellow tinge. Several cycles of applying *B. subtilis* were required over 3-4 days to achieve a total dry weight of 630 milligrams (250 milligrams clean cloth plus 380 milligrams of dry *B. subtilis* cells). This represents a contaminated cloth that is now 60% bacteria by weight and 40% cotton cloth. As referenced earlier, the radiocarbon laboratories estimate that an organic artifact would have to have a contamination level of 60% to achieve a radiocarbon date of the first century. Additional applications of bacteria and growth cycles could achieve much higher concentrations of bacteria relative to the mass of the cloth. The physical appearance of the contaminated cloth is of a yellowed "aged" fabric (see photo) quite similar to the fabric of the Shroud of Turin. The "yellowing" is due at a minimum to the oxidation of unsaturated fatty acids upon direct exposure to oxygen and drying. These results are easily duplicated by anyone with the tools to grow bacteria on cellulose cloth surfaces.

It is understood that this simulation of microbial contamination is not reflective of the true complexity associated with contamination of the Shroud of Turin. This contamination event required only a few days. Imagine the opportunities for microbial contamination that could be achieved naturally in several months or several years not to mention centuries or several thousand years. The variety of species inhabiting the Shroud microbiome must be in tens of thousands and the number of individual bacterial and fungal cells must be in the billions to trillions. The microbes are not consuming the linen fabric but simply using it as a niche or residence. The only microbes that would destroy the Shroud would be bacteria and fungi found in the soil that produce the enzyme cellulase, which would break down the cellulose linen fabric. Since we still have the Shroud, we can assume that no cellulase producing microbes have been able to establish long term residence on the Shroud linen.

Figure A

In the photograph above, the uncoated one on the left weighs 250 milligrams (dry weight). The one on the right was originally 250 milligrams but 380 milligrams of bacteria were added to give a final dry weight of 630 milligrams. Thus, 40% of the weight is due to the original untreated cloth and 60% of the final weight is due to added bacteria.

1.9 Do it Yourself – Getting Your Microbiome to Create a Shroud-type Image

Procedure for Creating Microbial Photographs from Skin

Isolate *Staphylococcus epidermidis* from skin using blood agar plates
↓
Grow *S. epidermidis* on the surface of multiple blood agar plates
↓
With a swab thoroughly cover the surface of the hand with bacteria
↓
This will be one million or more bacteria per square centimeter
↓
After drying, add water and smooth evenly over the hand
↓
Cover hand with a moistened linen cloth and dry completely
↓
Keep hand still or blurring will occur
↓
Once linen is thoroughly dry, remove slowly from back of hand
↓
Compare the positive yellow image with a photographic negative image

Seeing for yourself is one thing but how do you "do it yourself?" "Seeing" for ourselves concerned microbial contamination. "Do it yourself" relates to replication of an image on linen material using the same microbial action that gave rise to the Shroud image. How do you do this? Very simply. With specific precautions, you recreate on parts of your body or on that of a volunteer, conditions similar to those operative on a wounded corpse placed in a shroud [CAUTION: The reader is warned not to attempt this procedure on the body because of its potential risks]. For instance, let's say you want to try this out on your hand. In this case, you apply the bacteria associated with skin on your hand and place a linen cloth on your hand for a period of time. Eventually a straw-colored image of your hand appears on the cloth that is functionally similar to the Shroud image. If you take a photograph of the image, it looks much like what you see in photographs of the Shroud.

As before, we describe replicable experiments performed by SJM this time demonstrating microbial image formation. It should be noted that the images will vary depending on the particular situation involving the subject. The Man of the Shroud underwent unique forms of physical torture and mutilation which cannot be precisely replicated so no image formed from subsequent replications will match what we see on the Shroud. Nevertheless, images formed from the relevant kinds of microbial action on human subjects will manifest eerie resemblances to the Shroud image.

The rest of this chapter as well as the chapter which follows describe in some detail the scientific experiments carried out by co-author SJM. As such it necessarily includes technical terminology and scientific specifications and narratives. Overviews of these experiments and their implications are described throughout the book. **Consequently, readers who are not inclined to wade through such scientific details can skip the sections below as well as the next chapter (1.10).**

Microbial Image Formation Case Study
by Stephen J. Mattingly (SJM)

Note – the Figures/photos referenced here are found in Appendix 1.

What part of the human body should be used in an attempt to create a reproduction of the Shroud?

The first impression I had of seeing the Shroud first-hand at the 2000 Shroud of Turin Exposition in Turin, Italy was of a straw-yellow color of a man with evidence of many body wounds. It was an extremely haunting and sad experience. The source of the straw-yellow color has never been explained scientifically. In a later question, I will demonstrate the origin of the color and its implications for those who found the burial wrappings in the deserted tomb. Another feature is that the image must be mainly on the surface of the linen with little penetration. This indicated that the image matter must be of high molecular weight and not subject to diffusion through the fabric.

Another exceedingly important feature is that a photographic negative image must show greatly enhanced detail compared to the original positive straw-yellow image.

In attempting to create an image in the laboratory, I have had suggestions on what I should use as a model, ranging from a live nude body to a recently deceased corpse. I am uncomfortable with either suggestion. Instead I decided to use my hand and face as models for creating images. I chose my hand because of a fairly flat surface with little fatty tissue under the skin. The bones of my hand can be seen to some degree under the skin and would represent a framework for the linen. My face would present a much more difficult challenge due to its various angles and slopes as well as the problems associated with the need to breathe. In addition, my cheeks and other areas of my face have fatty deposits under the skin which would not permit the linen to lie directly over a bony framework. Features of my face would have a more rounded appearance than we see with the face of the Man of the Shroud. Once I have discussed the results obtained with my hand and face, I will explain why we do not see some of the same results evident in the image on the Shroud of Turin.

How are skin bacteria grown to create an image?

Before I present specific details on how to produce an image, it is necessary to address some important safety precautions. Since the hand is the most convenient to image, it is essential that it does **not** have any cuts or wounds. It is possible to cause an infection with the subject's own bacteria, because we will be coating the hand with *Staphylococcus epidermidis* at levels never experienced. Massive amounts of bacteria at a wound site could invade the blood stream and cause bacteremia or septicemia and be dispersed throughout the body.

In order to isolate *S. epidermidis* from my hand, I ran a sterile swab over the back of my hand as shown in Fig. 1. I then swabbed it over a trypticase soy agar plate containing 5% whole sheep blood as shown in Fig. 2. These plates are commercially available and are in routine use in hospital clinical laboratories. I then incubated the plate at body temperature (98.6° F) for 18-24 hours. A number of white colonies appeared on the plate and I selected one colony for the study. I examined it under the microscope after performing a gram stain and found gram positive cocci in clusters. I then performed a catalase and a coagulase test, which are routine procedures to distinguish between *Staphylococcus epidermidis* and *Staphylococcus aureus*. Both bacteria are catalase positive, which distinguishes them from the streptococci. The coagulase test was negative for my isolate identifying it as *S. epidermidis*. *S. aureus* is a human pathogen and is coagulase positive. It is not part of the normal bacterial flora of the skin. Thus, these procedures are critical to identify the correct *Staphylococcus* to be used to create the image. Use of *S. aureus* could cause a life-threatening infection. These laboratory procedures are available in any recent medical microbiology textbook or clinical laboratory manual.

I then swabbed my isolated colony over the surface of several other blood agar plates and incubated them 98.6°F for another 18-24 hours (Fig. 3). I then coated the surface of my left hand with bacteria from the blood agar plates (Fig. 4). I ran very rough estimations of the number of bacteria applied to my skin by doing colony counts. The results showed that approximately 20 million bacteria/cm^2 were applied, although because of the clumping of staphylococci the number could be off by a factor of 10 or more. After allowing the bacteria to dry on my hand, the left hand had a only a slight sheen compared to the right (Fig. 5)

These results demonstrate that the skin of a body coated with a massive overgrowth of bacteria is virtually indistinguishable from a normal uncoated body. As we will see, the presence of water, however, clearly has a profound effect on their appearance and the physical properties of skin surfaces harboring high levels of *S. epidermidis*.

What is the role of water on image formation?

> *One tends to forget that all the cells of a corpse continue to live, each one on its own, those of the skin like the others, and they die individually after different lengths of time....If the higher-grade and the nervous cells are the most fragile, yet the others last for some time; total death only sets in with putrefaction....On the other hand all the wounds, all the abrasions with which the body was covered continued to exude a more or less infected lymph as when it was still alive, but in liquid form.*

> *The result of all this was that the body was bathed in a watery atmosphere, which made all the clots on the skin and in the various wounds damp once more.* Pierre Barbet, *A Doctor at Calvary*

The greatest surprise to me in this study was the unexpected role that water played. My intention was to coat my hand with unusually high levels of my own skin bacteria and after drying wash my hand thoroughly with water as I anticipated the followers of Jesus would have done with his body before burial. I began by running my coated hand under water. Instantly, my hand became coated with a thick, gelatinous viscous material (Fig. 6). I realized that I was visualizing for the first time the reality of the effect of the scourging and crown of thorns on the skin surface of the Man of the Shroud. Adding blood to this material, you can see the image of the suffering Jesus in great detail.

What was the nature of this dramatic effect with water? I have done research for 33 years on the extracellular polysaccharides of several bacteria and I am very familiar with the properties of these polymers. Perhaps, the largest and most viscous exopolysaccharide that I studied was the alginic acid polysaccharide produced by *Pseudomonas aeruginosa*. It plays the predominant role in damaging the lungs of cystic fibrosis patients.

I was aware of the exopolysaccharide produced by *Staphylococcus epidermidis*, but not on its ability to absorb large quantities of water. Bacteria that do not produce exopolysaccharides do not respond to water in this manner. In considering this effect with *S. epidermidis*, I better understand the normal role of this exopolysaccharide in maintaining hydrated skin cells by absorbing moisture from the air. I have not seen this effect reported in the medical literature.

In an attempt to clean my hand of this viscous material, I moistened a linen cloth and rubbed it over my hand repeatedly without being able to remove all the coating. What I had accomplished was to spread the material evenly over my hand as the followers of Jesus would have done in attempting to clean his body. In the process of cleaning my hand with water, I became aware that the viscous material was extremely sticky. Undoubtedly the glue-like polysaccharide plays a major role in cementing *S. epidermidis* to our skin epithelial cells. It also will have profound effects on "gluing the linen" to the body surface.

As indicated before, a series of events must occur in a precise manner to result in the formation of the image on the Shroud of Turin. As these events increase in number, the likelihood that this is indeed the burial Shroud of Jesus also increases.

What features of linen contribute to image formation?

The Shroud linen is known to have a 3:1 herringbone weave pattern and was considered to be an expensive textile in the first century. My choice of linen, in contrast, was from Wal-Mart. The linen had a weave pattern sufficiently dense that allowed it to lie on my hand showing the various contours. The linen was pre-washed to remove any chemicals or debris. To begin the imaging of my hand, I laid a slightly damp linen cloth without applying any pressure on the back of my *Staphylococcus epidermidis* coated left hand (Fig. 7). (A damp cloth was used because the linen that wrapped Jesus' body would have been damp given both that he was washed with water and a recently deceased body oozes water at the surface as skin cells begin to die.) The drying process required about 30 minutes and considerable effort was made not to move my hand during this time. As you will see in subsequent questions, the imaging process has the basic characteristics of photography.

The photographic emulsion was the coating of bacterial cells with their water-seeking exopolysaccharide, while the film was the linen itself. Any movement of my hand in the drying process will create a final blurring of the image. This was certainly not an issue with the deceased Man of the Shroud.

During the process of drying, the linen was drawn tightly to my skin and I could feel the fabric increasing its binding force over time. The only thing I can compare this to is having a sweaty tee-shirt dry on your back after vigorous exercise. However, in this case I could feel the hairs on the back of my hand being pulled and stretched. After complete drying, the basic features of my hand are evident in Fig. 8. The fabric actually penetrated into the folds of the back of my hand and tightly around my fingers. It is apparent that the linen is seeking the bones of my hand as a framework for final attachment. When we consider the nature of the crucified body of Jesus at death, we will understand why his image on the Shroud was a one-time occurrence and can never be adequately reproduced.

Nevertheless, these results provide extremely basic information for understanding how the image on the Shroud of Turin formed. The most important characteristic of the linen is its ability to transfer any water from the linen itself or even indirectly from the air through the linen to the bacterial exopolysaccharide "emulsion" coated on the back of the hand. We will now examine the time required for image formation as we remove the linen from my hand.

When did evidence of an image appear?

I allowed the linen on my hand to dry thoroughly (approximately 30 min) before I began to slowly remove it. As I pulled up on the edge of the linen I found that it lifted up the skin on my hand and considerable effort was exerted to separate it from my skin. (Fig. 9) As I looked under the linen, I immediately saw traces of a yellow color forming (Fig. 9). Thus the image formed "instantaneously" upon exposure to air (oxygen). This event is an important clue to the type of biomolecules that immediately forms a straw-yellow color on exposure to oxygen resulting in oxidation of the material.

As I continued to pull the linen from my skin, I noticed that it was pulling out a number of hairs from the back of my hand causing some discomfort. I hope this gives the reader some idea about how tightly the linen was sealed to my hand. As I continued pulling the linen off my hand, the image was becoming more defined and I could see how it formed around my fingers (Fig. 10). When it was completely removed I laid it down to watch the color continue to develop. In the meantime, I noticed the effect that the drying process had on the appearance of the back of my hand (Fig. 11). Compared to my right hand, my left hand had "aged" considerably based on the obvious increase in wrinkles. Amazingly, the water-seeking exopolysaccharides of *Staphylococcus epidermidis* had withdrawn water from the tissues on the back of my hand. We have all witnessed wrinkled fingers if we keep our hands in bath water or dish water too long. However, this was a most unusual occurrence with drying linen and it gives us additional insight into the mechanism of image formation on the Shroud.

Once the body of the Man of the Shroud had been wrapped with linen and placed in the tomb, any remaining water on the body or the linen would have been absorbed by the *S. epidermidis* exopolysaccharides as shown with my hand. Additional moisture would have been drawn from the atmosphere and the skin tissues of the deceased. As drying continued, a stronger pull from the exopolysaccharides would draw the Shroud linen tighter and tighter to the surface of the body. This would serve, therefore, to create the exquisite detail we see in the image on the Shroud of Turin.

How do the positive and negative hand images compare?

The positive straw-yellow image and the basic definition of my left hand are readily apparent in Fig. 12 when the linen is turned over. The reader should keep in mind that the underside of the linen is a mirror image of my left hand. There appeared to be some blurring along my index finger as well as near the backside of my hand undoubtedly caused by movement during the drying process. My fingers and thumb are well defined as are all features of my hand including my wedding ring. However, in comparison to the negative image in Fig. 13, the remarkable detail as seen in the Shroud of Turin is striking. Some blurring can also be seen around the thumb as well as the index finger. My first reaction was of amazement at the clarity and detail of the negative image.

This experiment has been replicated many times by my students. It is straightforward and scientifically sound. With the precautions I stated earlier, I am sure others can produce their hand images, some with even greater detail than mine. I will caution you that holding the hand still for up to 30 minutes while the linen dries is not easy. [TO REPEAT, PLEASE DO NOT ATTEMPT TO REPLICATE THIS EXPERIMENT ON YOUR OWN]

How does this image compare to the detail in the Shroud of Turin? Of course, no live imaging will ever be as defined as seen on the Shroud. The obvious reason is that dead bodies do not move. The relative dampness of the linen can leave an outline around the fingers and hand that is not seen on the Shroud. This suggests to me that much of the moisture was drawn from the body over a period of time and not from the washing procedure used to clean the body. Thus, this is another reason why this was a one-time occurrence that can never be repeated. We will never know the exact condition of the body at death as well as environmental factors during preparation for burial or the conditions in the cave after burial. However, the **process** by which the image formed can be duplicated by anyone with access to a microbiology laboratory. I have presented these findings at scientific meetings of my peers and have found acceptance and encouragement. Non-microbiologist critics have focused instead on my inability to exactly replicate the precise kind of image found on the Shroud. The criticism is misguided because it is an obvious impossibility for us now to replicate the exact circumstances faced by the Man of the Shroud. Similitude will have to do when exactitude is impossible.

How do the positive and negative face images compare?

Producing a face image is not as direct as a hand image. For one thing, I did not have the facilities in my laboratory to recline in a comfortable position for 30 minutes or more. I was also concerned that someone would find me and I would not be able to reply. So, I took the experiment home where I would have the assistance of my wife. She spread *Staphylococcus epidermidis* as evenly as possible over my face, avoiding the area around my eyes. **Caution: do not attempt this procedure.** Unfortunately, the odor of the bacteria caused her to become nauseous and I had to take the experiment outdoors. I reclined flat in a lounge chair and waited until the linen was dry to the touch. Another problem with imaging the face is the need to breathe. Thus, the linen near my nose was never able to seal properly. As I removed the tightly bound linen from my face, I became aware of large black birds sitting quietly along the gutter of the house and the nearby fence. I never heard them approach or land. I assume these birds sensed the "smell of death" and saw me as their next meal drawn by the odor of the bacteria.

The positive straw-yellow image (Fig. 14) clearly shows my mouth, lips, nose, eyebrows and cheek areas in some detail. As indicated earlier, the area around my eyes was avoided and appears light as does the area under my nose. As with the Shroud of Turin, the negative image (Fig. 15) shows much greater detail of all areas. I easily recognized my facial features.

Another limitation in imaging the living is the rounding of the image near the edges. This is due to hydrated tissues surrounding the supporting bone structure. This is not seen with the Man of the Shroud, who has quite angular features. In the case of my hand, there was limited tissue over the bones and access to bone was easier. However, my cheeks are quite full and hydrated limiting access to the cheek bone. Another problem I will deal with in a later question concerns the formation of the image in an area that probably did not make contact with the body.

What is the source of the straw-yellow color seen in the hand and face images and the image of the Shroud of Turin?

As indicated in an earlier question, the instantaneous straw-yellow color development when my hand image was exposed to air (oxygen) was a clue to its molecular nature. The only biological molecules which I am aware of that form this color when exposed to oxygen are pigments of various types and unsaturated fatty acids. *Staphylococcus epidermidis* does have a pigment(s) that can produce a yellow color and could very well contribute to the color development. However, it is unknown if the color development is produced only under conditions where oxygen is present. It is known that unsaturated fatty acids found in the membranes of *S. epidermidis* and any other cell, including human epithelial cells, produce a straw-yellow color when exposed to oxygen. A familiar occurrence of this yellowing is "ring around the collar" found inside the collar of shirts, most notably white shirts. This yellow ring is due to the bacterial and skins cells being deposited on the fabric with their membrane components including unsaturated fatty acids. This occurs over a period of time as the dead cells accumulate on the collar.

To examine the source of the straw-yellow color experimentally, I dried (lyophilized) a 1 gram culture of my *S. epidermidis*. The cells were extracted by the Folch procedure with chloroform/methanol (2/1) and, after agitation, the two phases separated. The chloroform phase contained water-insoluble unsaturated fatty acids and the methanol phase contained water soluble molecules. I spotted a white filter paper with a sample from each phase and allowed it to dry. As shown in Fig. 16, the yellow color was present in both the chloroform and methanol phases. The left side is the methanol phase and the right is the chloroform phase. Thus the unsaturated long chain fatty acids specifically contribute to the yellow color, since they will only be found in the chloroform phase. The methanol phase included water-soluble pigments and degradation products of fatty acids resulting from the oxidation in air. More research would be needed to completely identify all the material resulting in color development in the presence of oxygen. However, at a minimum, unsaturated, long chain fatty acids explain the formation of the straw-yellow color of the images.

Thus the image on the Shroud is both an oil and a water color "painting" made by the remains of a crucified body. The water color "painting" would be subject to the changing humidities and slowly dissipate over time unlike the oil "painting" which would be resistant to these factors.

How can an image form without direct contact with skin?

Because of all the contours of the human form, when covering the front and back of the body with linen as in the Shroud, there are likely to be areas of skin that would not make direct contact with the cloth. However, as described earlier, the hydrostatic pull of water from the body for the exopolysaccharides on the linen might be sufficient to stretch the linen until it made contact with all skin surfaces of the body. That certainly seemed to be the case with my hand and face images.

Nevertheless, I explored another possibility looking at the ability of *Staphylococcus epidermidis* to move over a linen surface on a blood agar plate. I streaked one end of the plate with *S. epidermidis* and allowed it to grow overnight (Fig. 17). The next day I placed a piece of slightly damp linen at a right angle next to the streak and allowed the bacteria to grow overnight at room temperature (Fig. 18). *S. epidermidis* does not have the ability to migrate over the surface of the blood agar plate (BAP) as indicated by the streak. It is restricted to the general area in which it was originally inoculated. The reason for this is that *S. epidermidis* does not have appendages for locomotion (called flagella) which allow bacteria, such as *Proteus*, to swarm over the surface of the BAP. However, please observe in Fig. 20 that the linen is now coated with a confluent growth of *S. epidermidis* which is restricted to the linen area. The explanation is that the linen maintained a moist surface and allowed a diffusion of nutrients from the BAP and subsequent growth of *S. epidermidis*. The linen pulled water from the BAP along with the dissolved nutrients.

Thus, diffusion of water and nutrients from an area in contact with skin to an area of linen not in contact would be analogous to the experiment described above. With these conditions in place, *S. epidermidis* would grow into the area not in contact with the skin and complete the image. These results demonstrate that even with the apparent inability to migrate over a surface, environmental conditions can exist that allow bacteria to disseminate and extend into areas that normally would be restrictive.

The role of *Staphylococcus epidermidis* in image formation is summarized by the painting (Figure 19) by Nancy Mattingly of the Man of the Shroud using only a mixture of *S. epidermidis* and water.

What other macromolecules can contribute to long-term image formation?

Any biological material made of carbon including the Shroud of Turin will oxidize and disintegrate over time. The linen fabric of the Shroud is composed of cellulose which is a polymer of glucose linked together in long chains with occasional branching. As I have demonstrated in the extraction of *Staphylococcus epidermidis* cells, the instantaneous yellowing of my images was due to water-insoluble unsaturated fatty acids in the chloroform phase and possibly their degradation products and a pigment(s) in the water-soluble methanol phase.

In addition to the instantaneous oxidation associated with unsaturated fatty acids, the other deposited material on the image transferred from the skin would very slowly transition through a yellowing phase which would include the exopolysaccharide capsule along with the cell wall peptidoglycan. This structure is identical to cellulose except that it has repeating molecules of N-acetylglucosamine instead of glucose and alternating N-acetylglucosamine with phosphoenolpyruvate at C3 (muramic acid) attached to a tetrapeptide. Proteins, ribonucleic acid (RNA), deoxyribonucleic acid (DNA) and a variety of other smaller molecules would also oxidize and yellow over an extended period of time. These molecules are much less reactive to oxygen than the unsaturated fatty acids and pigments. You might have noticed how long it takes pages (composed of cellulose) in a book to turn yellow. Centuries may go by before any noticeable yellowing occurs.

In the case of the Shroud, there is a greater density of highly reactive material deposited in the image area than in the non-image area. Thus, the image stands out in contrast to the non-image area. Assuming care was exercised in the handling of the Shroud over the centuries, the image area would likely show deeper color development over time in contrast to the non-image area because of the more reactive molecules in the image.

1.10 Letting Sleeping Bugs Lie – Demonstrating the 1988 Tests' Failure to Decontaminate

We have seen "for ourselves" that microbial contamination is an indisputable, instantly verifiable fact of the physical world. There is "proof of life" everywhere on earth *if we know how to look for it*. Again, by "doing" it ourselves, we can confirm that under certain conditions – conditions precisely reported in the Passion narratives – microbial action can create an image of the kind we find on the Shroud. Both these procedures meet the scientific criterion of verifiability. Namely, unless a claim can be verified through an experiment that is, in principle, replicable by anyone, it is not scientifically acceptable. Microbial contamination and microbial image formation can both be replicated – by anyone. Hence, they conclusively establish the Shroud's embedded microbiome.

We show below through yet another replicable procedure that the cleaning processes used for the 1988 C14 dating (its published version is also given in this chapter) could not and did not rid the Shroud of its microbiota. And thus its dating results were fatally flawed. In fact, there was no way to effectively de-contaminate the Shroud samples without destroying them.

First, we will review the replicable experiment (immediately below) that demonstrates the inability to microbially decontaminate samples using the C14 team's cleaning procedures. Following that we will do a comparative review of the cleaning procedures listed by the C14 team.

Stephen J. Mattingly and Roy Abraham Varghese

Decontamination Case Study – Effect of Various Treatments used on the Shroud of Turin to Clean 100% Cotton Cloth Artificially Contaminated with Known levels of Bacteria

A variety of techniques were used to "clean" the Shroud linen before determining the radiocarbon date of 1260-1390 AD. The harshest treatment of the linen utilized 1 M HCl at 80 C for 2 hours, rinsing with water, followed by 1 M NaOH at 80 C for 2 hours, rinsing with water, and again with 1 M HCl at 80 C for 2 hours and rinsing with water. No studies were reported on the effects of these treatments on the linen fabric. In addition, in the 1988 test, no chemical analysis was conducted to determine the actual level of contamination before and after treatment. Since linen is primarily cellulose (glucose) in composition, then a glucose content of 80-90% or more would assure the purity of the samples and provide confidence in the radiocarbon dating results. In the present study, the harshest cleaning technique (as described above) was utilized to first determine the effect on clean 100% cotton cloth and then on bacterial coated cloth. Three 250 mg samples of clean cloth were treated by the harshest procedure described above with less than a 2 milligram difference in the dried samples after treatment. The cleaned samples are indicated on the left in Fig B. Importantly, these results demonstrated that the harshest cleaning treatment used by the radiocarbon laboratories had no effect on the structural integrity and solubilization of the 100% cotton fabric itself and presumably also on the Shroud of Turin linen.

The cleaned fabric (250 milligram) was then contaminated with the aerobic, Gram-positive, spore-forming, air contaminant, *Bacillus subtilis*, over a four day period. *B. subtilis* was grown on nutrient agar plates overnight at 35 C and then colonies were picked and mixed with 1-2 ml of Thioglycollate broth and applied to the fabric and allowed to continue to grow at 35 C overnight. The next day the contaminated cloth was dried and weighed. Additional amounts of bacteria were added and grown until the weight of the cloth plus added bacteria reached a dry weight mass of 630 milligrams. This represents a ratio of 60% bacterial contamination to 40% clean linen by dry weight. The contaminated sample is indicated in the middle in the Figure below.

The contaminated cloth had become discolored, was stiff and had a plastic-like texture. Much higher levels of bacteria could be added further decreasing the relative contribution of the linen component. These results demonstrate that bacteria can add considerable weight to an object without changing the shape or size. A third sample was contaminated with bacteria to achieve 60% bacterial contamination as before and was subjected to the harshest cleaning treatment as indicated above. After treatment, there was a loss of 60 milligrams of dry weight which represents 16% of the added bacterial mass. The treated sample is indicated on the right in the Figure below.

The loss of cell mass in the acid:base:acid cleaning study described above can be attributed to the solubilization of DNA, RNA, fatty acids, cytoplasmic membrane components, loosely associated proteins on the exterior of the cell, as well as intracellular molecules that are released upon the solubilization of the cytoplasmic membrane. Thus, cleaning of the Shroud of Turin by the harshest treatments reported by the radiocarbon laboratories would leave behind insoluble macromolecular carbon-containing protein and peptidoglycan material that would markedly affect the radiocarbon dating results. Of the two major macromolecules remaining on the cloth, the most important that cannot be removed by any treatment employed by the radiocarbon laboratories is the cell wall peptidoglycan. This structure has the same beta-1,4-glucose-linked backbone as cellulose, the plant based polymer of cotton and linen fabrics. This is also the same polymer found in the cell wall chitin of fungal contaminants. Thus the bacterial peptidoglycan left behind by any of the treatments would not be recognized as different from the linen cellulose by most analytical or visual techniques. Purified peptidoglycan is white as is cellulose. The only means of definitely identifying it as of bacterial origin would be a complete chemical analysis showing the presence of muramic acid, a glucose molecule with a three carbon phosphoenolpyruvate added at carbon 3 of glucose. In addition, an N-acetyl group is added at carbon 2 of glucose. N-acetylmuramic acid and N-acetylglucosamine alternate in long strands of peptidoglycan connected by short peptide bridges composed of several different amino acids. Muramic acid is unique to the bacterial world and is a definitive marker for the presence of bacteria. Thus, the presence of muramic acid in a sample to be carbon dated would indicate bacterial contamination and samples would need to be subjected to additional cleaning before carbon dating. Likewise, the other major component left behind by all cleaning techniques employed by the radiocarbon laboratories is protein, which is composed of carbon-based amino acids linked through highly resistant covalent peptide bonds. **In order to remove and solubilize both proteins and peptidoglycan macromolecular polymers, the standard hydrolytic procedure requires 6 M HCl at 100 C for 24 hours.**[1] This treatment would obviously also destroy the Shroud of Turin linen fabric as well as any cellulose-based fabric. **Thus, radiocarbon dating of any ancient surface-exposed object is a complex undertaking requiring definitive chemical analysis to assure purity of the sample and reliability of the dating.**

Figure B

Comparison of three pieces of 100% cotton cloth each weighing approximately 250 mg. The left cloth was treated with 1M HCl for 2 hr at 80C and rinsed in water, then 1 M NaOH for 2 hr at 80 C and rinsed in water, followed by 1M HCl for 2 hr at 80C and rinsed in water and dried to constant weight. There was no loss in weight. The middle cloth and the cloth on the right had a total of 380 mg (dry weight) of Bacillus subtilis added which is 60% of the total weight of the two cloths. The right cloth was then treated with the HCl and the NaOH procedures as described above. A total of 60 mg of bacterial dry weight was solubilized leaving 84% dry weight of the original added bacteria still on the cloth.

Carbon 14 Test Cleaning Procedures

But what of the 1988 Carbon 14 dating of the Shroud? We have already noted that the dating team was handicapped by the absence of microbiologists who could consider the Shroud microbiome. This oversight was all but inevitable given that the dating specialists of the time were simply not aware of the all-pervasive presence of microbial agents. The crucial issue in evaluating the efficacy of this initiative is determining whether or not the investigators were successful in decontaminating the Shroud sample prior to the dating study. If the decontamination was ineffective then the test has no bearing on the age of the sample.

The official publication of the results of the 1988 study detailed the cleaning procedures used:

> The Arizona group split each sample into four subsamples. One pair of subsamples from each textile was treated with dilute HCL, dilute NaOH and again in acid, with rinsing in between (method a). The second pair of subsamples was treated with a commercial detergent (1.5% SDS), distilled water, 0.1% HCL and another detergent (1.5% triton X-100); they were then submitted to a Soxhlet extraction with ethanol for 60 min and washed with distilled water at 70° C in an ultrasonic bath (method b).

> The Oxford group divided the precleaned sample into three. Each subsample was treated with 1M HCL (80° C for 2h), 1M NaOH (80° C for 2 h) and again in acid, with rinsing in between. Two of the three samples were then bleached in NaOCL (2.5% at pH-3 for 30 min).

> The Zurich group first split each ultrasonically cleaned sample in half, with the treatment of the second set of samples being deferred until the radiocarbon measurements on the first set

had been completed. The first set of samples was further subdivided into three portions. One-third received no further treatment, one-third was submitted to a weak treatment with 0.5% HCL (room temperature), 0.25% NaOH (room temperature) and again in acid, with rinsing in between. The final third was given a strong treatment, using the same procedure except that hot (80° C) 5% HCL and 2.5% NaOH were used. After the first set of measurements revealed no evidence of contamination, the second set was split into two portions, to which the weak and strong chemical treatments were applied.[2]

These procedures involved ultrasonic treatment to remove any bound or associated debris that was not an integral part of the linen. In the *Bacillus* experiment described previously, ultrasonic treatment might release some of the bacterial mass since it has not had time to weave itself through the fabric. However, after months and years of growth, it along with other bacteria and fungi would be found throughout the fabric and within the lumen of fibrils. The various acids, bases, oxidizing agents and solvents along with temperatures employed by the radiocarbon laboratories failed to bring about the activation energy needed to break covalent glycosidic and peptide bonds found in bacterial peptidoglycan and glycosidic bonds in fungal chitin. Because they did not destroy the Shroud of Turin linen, we know that the activation energy needed to break glycosidic bonds in plant cellulose was also not achieved.

In brief, since the sample was still contaminated, the C14 study and its results were null and void with respect to the Shroud's age.

Spores

Another neglected area relevant to decontamination is the bacterial endospore. The *Bacillus* genus I used in the experiments reported here produces endospores. There are many species and they are found in both the soil and air. Each vegetative cell can produce one spore (endospore). These endospores are the most resistant life form on planet earth and have been shown to survive millions of years. The *Bacillus* produces endospores when growing conditions become unfavorable. When the right conditions exist, the spores will germinate into vegetative cells and grow again. They do not produce cellulase so they will not damage the Shroud linen. On something as large as the Shroud, there are likely to be billions of spores present.

As noted, the dating investigators did not use the cleaning procedures required to decontaminate the Shroud at a microbial level. In addition, the procedures they used would not have dissolved and eliminated spores. Spore resistance even to radiation, heat and chemicals is well-known.[3]

1.11 FAQs – Critiques of the Microbial Paradigm and Responses

When SJM first introduced a prototype of the embedded microbiome paradigm in the Nineties, the fiercest critics were Shroud "believers." Some of these antagonists were irate because the image formation mechanism did not include a supernatural agency. They forget that Jesus himself said, "if they keep silent, the stones will cry out!" *Luke* 19:40. To be sure, Shroud skeptics were also hostile to the microbial approach but several of them were open to its possibilities. Curiously, formal presentations of this paradigm to professional groups like the American Society for Microbiology and others met with no objections!

Almost without exception, the critics are not themselves biologists. And the concerns they raise often highlight their ignorance of elementary microbiology. Their critiques illustrate the dangers of authorities from one discipline pronouncing on specialized topics in another – in this instance, physicists, chemists, journalists, et. al. holding forth on the minutiae of microbiology.

In previous chapters, we have *demonstrated* that the main charge below made by critics is embarrassingly mistaken:

> The microbial world plays no appreciable role in the composition of ancient organic material; if at all it played a role, this would be detectable but no such role is detected in the Shroud and, in any case, every conceivable "cleaning" method was applied to it prior to the radiocarbon testing; and if the microbial role was to be significant from a dating standpoint,

then it should increase the mass of the material two-fold. More than half the Shroud should be composed of "foreign matter."

Here we will respond to specific critiques of various elements of the embedded microbiome paradigm. These may be categorized under the following heads (freeform critics, of course, touch on more than one category):

- Microbial contamination was not possible
- The Carbon 14 labs' cleaning procedures would have removed microbial contaminants
- The contamination required to change the carbon-dating would require that half or more of the Shroud's weight is made up of contaminants
- An image derived from microbes would look different from what we have on the Shroud.

Microbial contamination was not possible

The main proponents of this hapless hypothesis were two chemists, Raymond Rogers and Alan Alder. The substance of their critiques, unfortunately, is a mixture of outright error and impoverished microbiology. Creditably, both have made contributions in their areas of expertise in Shroud studies. But their comments on microbial contamination constitute a textbook case of basic errors made by non-specialists delving into domains that require a lifetime of research and training.

Rogers:

> Mattingly's postulation of an appreciable amount of slime/biopolymer requires photosynthetic aerobes. ... Appreciable photosynthesis would not be expected on the Shroud, because historically it was stored dry in a dark place. ... STURP used all of the protein spot tests . ... There was no protein in areas other than the blood flows.[1]

Alder:

> It should be noted that to corrupt the observed radiodate from a first century date to that reported requires about a 50% increase in the C14 mole fraction. This is a prodigious amount of bacterial metabolism. ... Where does all this energy for growth come from? Are the organisms photosynthetic? Where does the mass come from? Does this microorganism fix the nitrogen from air as required for its growth and metabolism? Where does it get its sulfur, phosphorus, and minerals from and to where have they disappeared?[2]

These critiques are based on somewhat elementary misconceptions about the microbial world and have already been addressed.

To reiterate what has already been said, contrary to Rogers' postulates on what it takes for a microbial presence, microorganisms are found *everywhere* on surfaces exposed to the environment. These microbes include bacteria and fungi and other microscopic eukaryotic life forms. There are no surfaces except in fires and volcanoes where they cannot be found. They are able to reproduce if certain environmental factors, such as temperature, nutrients, moisture, correct pH, trace elements are present. Therefore, to discount their presence on the linen surface of the Shroud is an elementary error.

How are microbes able to grow on the Shroud surface? And where, to quote Alder, does their "growth come from?" In addition to the microbial life forms continuously falling on the Shroud from the air, many people have left biological materials on the Shroud during the many centuries. These include epithelial cells and oils from the skin along with bacteria from the skin. These would provide a rich source of nutrients for other bacteria, such as proteins (amino acids), carbohydrates, lipids and fatty acids, DNA and RNA, The failure to detect these biochemical molecules using the methods cited by Rogers indicates a failure of the technology employed. The spot tests he mentions would only detect massive levels of the markers, such as proteins. A much better approach would be to do an automatic amino acid analysis which would indicate the presence of all the amino acids.

This would require the complete acid hydrolysis of the proteins, which unfortunately would also destroy the linen fabric itself. Intact bacterial and fungal cell walls would block access to any proteins within the cell. So Rogers' spot tests have no applicability or validity.

How can the linen cellulose be distinguished from bacterial and fungal components? Unfortunately it cannot be distinguished. As indicated in a previous chapter, the backbone of linen cellulose, bacterial cell wall peptidoglycan, and fungal cell chitin all have the same beta-1,4- glucose-linked polymer backbone. The only differences between these structures are small peptides found in bacterial peptidoglycan and n-acetylated amino groups on peptidoglycan and chitin. The only ways to detect their presence again would be an automatic amino acid analysis which would end up destroying the fabric. Any other analytical procedure, including spectrographic analyses would not reveal any difference, because they are the same polymer. Likewise, the harshest procedure used by the carbon dating laboratories, employing 1 M HCL, 1 M NaOH, 1 M HCL, individually for 2 hours at 80 C, did not solubilize the linen cellulose and thus could not solubilize and clean the bacterial cell wall peptidoglycan or the fungal cell wall chitin because they are the same structure. The only way chemically to clean the linen of bacterial peptidoglycan containing peptides would be a treatment of 6N HCL for 24 hours at 100 C, which would unfortunately also dissolve the linen cellulose. Completely purified dried plant cellulose, bacterial peptidoglycan, and fungal chitin all have the same white color so visual inspection would not reveal any differences.

The Carbon 14 labs' cleaning procedures would have removed microbial contaminants

This claim comes from the chemist Walter McCrone, well-known for his eccentric and evidence-less declamations about the Shroud:

> In the first place, any sample sent to any carbon-dating lab (and I have sent a large number of them) will be cleaned up very carefully before they determine the carbon date so that in the

Shroud case, there was pretty pure linen that was tested. I send them linen from paintings that has absorbed varnish, media, and pigments and is only about 30% by weight linen and they have to do a very good clean up job before it can be tested.[3]

McCrone wrote this in 1995. In 2008, Christopher Ramsey, the head of the Oxford lab which along with two others did the C14 testing, acknowledged that "There are also other possible types of contaminant, and it could be that one, or some combination of these, might mean that the Shroud is somewhat older than the radiocarbon date suggests."[4] More to our point here, AMS C14 testing pioneer Harry Gove observed in a TV interview that microbial contamination "of the linen fibrils could not have been removed even by the most stringent pretreatment cleaning process and would, definitely, skew the real age of the linen."[5]

The question can be decisively settled if we replicate the cleaning procedures used by the C14 labs with contaminated linen. This is precisely what SJM has done as described in the previous chapter. The results (as will be apparent to anyone who replicates the experiment) were unmistakable: the procedures used by the labs did not decontaminate the linen at a microbial level. Hence the C14 dates have no bearing on the actual age of the Shroud.

The contamination required to change the carbon-dating would require that half or more of the Shroud's mass is made up of contaminants

This brings us to the "most favored" critique: the contamination level required to "add" 1200+ years to the C14 date is implausible because it would mean that contaminants make up over half the mass of the Shroud. And this is absurd! This is the charge made by Alder and others cited below.

Shroud Center:

To skew the radiocarbon date based on a carbon dioxide source of C14 from the first century to the fourteenth century date proposed by the radiocarbon tests would essentially require a near doubling of the mass of the sample by the bioplastic contamination.[6]

Harry Gove:

As can readily be calculated, to change the radiocarbon age of the Shroud from the published date of 1325 AD to the first century, 74% of the Shroud sample supplied to the three laboratories would have to be modern carbon contamination, and only 26% original Shroud material. Visual inspection of the samples indicated that such a massive amount of modern carbonaceous contamination was extremely unlikely to be present either before or after the cleaning methods all three AMS laboratories employed.[7]

Robert Hedges, Oxford University:

If the shroud was originally 2,000 years old, but is contaminated by modern material to give a date of AD1250, the labs must have measured material contaminated by 60% modern, 20th-century biofilm. I find this incredible. It would be more biofilm than cellulose.[8]

Anyone who has read the previous chapters should be able to spot the obvious fallacies here. Yet again we have non-microbiologists treading on what looks like the solid ground of "common sense" only to sink swiftly into the quicksand of domain ignorance. The idea that high levels of contamination are not possible in ancient artifacts is demonstrably wrong. As it happens, we have provided replicable experiments definitively refuting this recurrent *non sequitur*.

The visual inspection described by Gove is irrelevant at best. You cannot look at *any* object and assume that it is appreciably free of microbial contamination. Every square millimeter of your skin is coated with a substantial layer of microorganisms. Along with your gut microbial flora, they contribute significantly to your body's overall cell numbers. Microbes, in fact, make up a significant percentage of the mass of living organisms. So it hardly strains credibility to say that they comprise more than 50 per cent of the constituents of a centuries-old piece of linen. In fact, what is not just implausible but scientifically incoherent is the notion that its microbial buildup would not have progressively increased with each passing century.

In the end, the only pertinent parameter is verifiable evidence. And what we have presented earlier is evidence that anyone who cares to can "see for themselves":

- the body imaged on the Shroud suffered from wounds resulting in bleeding;
- specific microbes are associated with the victim's skin and would start multiplying immediately;
- these and succeeding generations and varieties of microbes will inevitably constitute most of the "material" of the funerary linen that hosted them;
- the specific kind of microbial contamination involved with the Shroud can be replicated on linen samples today and such replication will show that over half the mass of the samples will eventually be made up of microbes.

The evidence for this degree of contamination was shared with carbon dating pioneer Harry Gove whose criticism was cited above. After extensive back-and-forth discussion on the matter over the years, SJM finally sent him two linen samples. One was uncontaminated and the weight was determined and included with the sample. A second sample with near the identical uncontaminated weight was coated with enough bacteria (previously killed by heat) to represent 60% of the dry weight of the linen sample. The contaminated sample was more yellow in color and had a stiffness similar to the Shroud. Sadly, Gove passed away some time after receiving the samples and we have no idea how he would have responded to the hard evidence now available. It is clear, nevertheless, that Gove was open to the possibility of microbial contamination skewing the C14 date for the Shroud. In fact, he went as far as to say, "This is not a crazy idea. A swing of 1,000 years would be a big change, but it's not wildly out of the question, and the issue needs to be resolved."[9] Gove's contributions to carbon-dating have been widely recognized and rightly applauded. But, personally speaking, SJM was just as impressed by his uncompromising commitment to following the evidence no matter where it led – the mark of a true scientist.

An image derived from microbes would look different from what we have on the Shroud.

Finally, there have been some critiques concerning the microbial mechanism of image formation.

> "The image is just too good to be true. ... Faces don't come out right for they have fat and muscle. He [Mattingly] argues that since the Shroud man's face came out clear he must have been skin and bone. He explains this by saying the man lost his muscle mass by dehydration and the loss of blood."[10]

Most critiques of the image formation either misunderstand what is being claimed or show ignorance of human anatomy. With respect to this critique, let us note the following: Water makes up 50-60 % of the human body. The lungs are 83% water and skeletal muscle about 70% water. The loss of massive amounts of water during crucifixion would be expected.

Of course, any attempt to microbially replicate an image today will not look exactly like the image on the Shroud. As we have said, that image was created under unique circumstances (crucifixion, etc.) and in specific environmental conditions and these cannot be duplicated. So, of course, an exact replication of the Shroud image is impossible.

Finale

To reiterate what we have said, there are two ways that the Shroud "carbon" could become younger over time. First, present day bacteria and fungi falling on the linen would add their new carbon to the linen. Second, as they grow, living heterotrophic bacteria and fungi incorporate some new carbon dioxide through the pyruvate carboxylase pathway. Autotrophic bacteria get all their carbon from present day carbon dioxide. These two pathways would result in an ever-younger Shroud linen. It should be clarified that these processes do not result in the Shroud linen being consumed and recycled. That would require the presence of cellulase which would damage the linen. There is no evidence of that kind of damage.

In closing it should be said that only microbiology furnishes a scientifically adequate explanation for the genesis of the Shroud image while also highlighting the impossibility of determining the Shroud's "age" using conventional technologies. But we cannot satisfy those who refuse to move beyond the old paradigms. Such obstinacy is nothing new in the history of science as Kuhn has documented: more often than not, the practitioners of "normal" science are unwilling to enter the thought-world of the revolutions that left them behind. So it has been with the new paradigm in Shroud studies. Those who cannot mentally make the paradigm-shift demanded by the microbial data will remain shut off from mainstream science in an intellectual ghetto of their own making. Like flat-earthers they can be consistent to a limited extent within their own thought-world. But the price they pay is exile from the world of data, evidence and reality.

Section II
Shroud in Toto

Hovering around all the manic debates over the Shroud is the specter of insanity. The image on the Shroud has launched a thousand ships of speculation sailing off in all directions at once: a medieval mastermind took a photograph *before* photography was invented; Leonardo da Vinci painted a self-portrait *before* he was born; a forger from the Middle Ages created the image using a type of plate glass first produced half a millennium *later*. Every possible explanation is invoked other than the one staring us in the face.

What began as an exercise in rational conjecture rooted in facts on the ground has exploded into a cacophony of competing fact-free theories. New hypotheses are announced virtually on a quarterly basis. Anything goes as long as the conclusion entails the Shroud not being a shroud. The only self-evident truth in Shroud studies is that all theories are created equal and are endowed by their creators with certain unreasonable rights, chief among which is the pursuit of sensation.

But enough is enough. It is time to leave the silly season behind us and return to sanity. We must turn on our critical faculties, apply the rules of evidence and, most importantly, look at the big picture. The most plausible hypothesis is one which has the widest explanatory scope and power while remaining true to what is obvious. It cannot contradict what is known scientifically but it is important to understand what this means: science deals strictly with the measurement of quantities: all else falls outside its scope: and sometimes measurement assumptions and even methods have to be revised thereby altering the resultant conclusions. Likewise, the hypothesis must be compatible with what is "known" historically but this does not mean that it must recreate every detail of the artifact's history (or museums could display no exhibits). Just as important, any hypothesis that flies in the face of obvious facts should be returned unopened to the sender. Once a hypothesis establishes its plausibility and is constantly consolidated with a procession of new data-points, then the burden of proof falls on its challengers.

What are the implications of this approach? For one, a hypothesis that cannot plausibly account for the origin of the Shroud image and such other salient features as the geographic varieties of pollen on it and the pre-medieval paintings of the Shroud should be shown the door.

A Panoply of Proof Points

The hypothesis propounded here is that the Shroud truly is the burial cloth of Jesus of Nazareth. This hypothesis is built on a panoply of proof points:

- hard facts,
- circumstantial evidence,
- corroborative testimony,
- authoritative attestations,
- observational data,
- compelling explanations.

The traditional claims about the Shroud being a burial cloth match what is obvious about it: it has bloodstains and the image of a victim who was crucified and scourged. The surface plausibility has been further substantiated by scientific studies of various kinds as well as historical and aesthetic analysis.

Those rejecting the traditional claims, on the other hand, have to explain away what is obvious while invariably resting their cases on mind-boggling and wholly implausible speculation. Data derived from certain scientific tools, C14 testing for instance, have been deployed to undermine the traditional claims. But such deployment is based on speculative inference from limited sub-sets of data rather than direct observation of all relevant evidence. In any case, such inferences do not even address the most basic questions such as accounting for the origin of the image. The C14 debacle illustrates the danger of resting a case entirely on speculative inference.

Religious Motive?

Now, while our hypothesis tallies with the traditional claims, we note that the Christian Faith does not require belief in the authenticity of the Shroud. The truth of the Christian world-vision is not determined by a judgment as to whether or not the Shroud is the burial cloth of Jesus. The resurrection of Jesus is held to be true for reasons other than the Shroud. The authenticity of the Shroud does give additional historical support for the crucifixion and also gives new meaning to the prophecy of the Suffering Servant. But the Faith is in no sense dependent on such support. The Shroud can be studied as a historical relic like Alexander the Great's Priene inscription or the Code of Hammurabi tablets. This is why there are agnostic and non-Christian researchers who accept the authenticity of the Shroud and Christians who do not. In short, this is not a quest driven by a religious agenda although some skeptics seem to be propelled by an anti-religious animus. Rather, it is motivated by the same human quest to explore and understand that gave us the great theories of science and the historical works of antiquity and modernity. We just want to know the truth and nothing but!

Our quest for the truth about the Shroud begins with the map that makes sense of what we see imprinted on it – the Passion and Tomb narratives of the New Testament.

2.1 Based on a True Story

The Shroud of Turin is no ordinary shroud. It is a shroud that tells a story. It seems, in fact, to be a real-time pictorial counterpart of the greatest story ever told – or, at least, the climactic part of that story, the Passion narratives. Beaten, scourged, crowned with thorns, forced to carry a cross, crucified, pierced with a spear, laid to rest in a linen shroud – this, according to the Gospels, is what happened to Jesus on the last day of his life. The Shroud shows us a man who underwent precisely this saga of suffering. Beyond the text-image correspondence of wounds, the Gospel accounts are additionally relevant for their references to Jesus' shroud.

So there are two kinds of relevant reference points:

the accounts of Jesus' suffering – *the Passion narratives*;
and the reports of his entombment in a linen shroud – *the Tomb narratives*.

As we see below, the very brevity and simplicity of the biblical accounts amplify the raw reality of what is being described.

The Passion Narratives

> Then they spat in his face and struck him, while some slapped him, saying, "Prophesy for us, Messiah: who is it that struck you?" *Matthew* 26: 67-68

> Then the soldiers of the governor took Jesus inside the praetorium and gathered the whole cohort around him. They stripped off his clothes and threw a scarlet military cloak about him. Weaving a crown out of thorns, they placed it on his head,

and a reed in his right hand. And kneeling before him, they mocked him, saying, "Hail, King of the Jews!" They spat upon him and took the reed and kept striking him on the head. And when they had mocked him, they stripped him of the cloak, dressed him in his own clothes, and led him off to crucify him. *Matthew* 27: 27-31

After they had crucified him, they divided his garments by casting lots; then they sat down and kept watch over him there. *Matthew* 27: 35

From noon onward, darkness came over the whole land until three in the afternoon. And about three o'clock Jesus cried out in a loud voice, *"Eli, Eli, lema sabachthani?"* which means, "My God, my God, why have you forsaken me?" ... But Jesus cried out again in a loud voice, and gave up his spirit. *Matthew* 27:45-46,50

Some began to spit on him. They blindfolded him and struck him and said to him, "Prophesy!" And the guards greeted him with blows. *Mark* 14:65

So Pilate, wishing to satisfy the crowd, released Barabbas to them and, after he had Jesus scourged, handed him over to be crucified. The soldiers led him away inside the palace, that is, the praetorium, and assembled the whole cohort. They clothed him in purple and, weaving a crown of thorns, placed it on him. They began to salute him with, "Hail, King of the Jews!" and kept striking his head with a reed and spitting upon him. They knelt before him in homage. And when they had mocked him, they stripped him of the purple cloak, dressed him in his own clothes, and led him out to crucify him. *Mark* 15:15-20

Then they crucified him and divided his garments by casting lots for them to see what each should take. ... At noon darkness came over the whole land until three in the afternoon. And at three o'clock Jesus cried out in a loud voice, *"Eloi,*

Eloi, lema sabachthani?" which is translated, "My God, my God, why have you forsaken me?" ... Jesus gave a loud cry and breathed his last. *Mark* 15:24, 33-34,37

The men who held Jesus in custody were ridiculing and beating him. They blindfolded him and questioned him, saying, "Prophesy! Who is it that struck you?" *Luke* 22:63-64

When they came to the place called the Skull, they crucified him and the criminals there, one on his right, the other on his left. *Luke* 23:33

It was now about noon and darkness came over the whole land until three in the afternoon because of an eclipse of the sun. Then the veil of the temple was torn down the middle. Jesus cried out in a loud voice, "Father, into your hands I commend my spirit"; and when he had said this he breathed his last. *Luke* 23:44-46

When he had said this, one of the temple guards standing there struck Jesus and said, "Is this the way you answer the high priest?" Jesus answered him, "If I have spoken wrongly, testify to the wrong; but if I have spoken rightly, why do you strike me?" *John* 18:22-23

Then Pilate took Jesus and had him scourged. And the soldiers wove a crown out of thorns and placed it on his head, and clothed him in a purple cloak, and they came to him and said, "Hail, King of the Jews!" And they struck him repeatedly. Once more Pilate went out and said to them, "Look, I am bringing him out to you, so that you may know that I find no guilt in him." So Jesus came out, wearing the crown of thorns and the purple cloak. And he said to them, "Behold, the man!" ... Then he handed him over to them to be crucified. So they took Jesus, and carrying the cross himself he went out to what is called the Place of the Skull, in Hebrew, Golgotha. There they crucified him, and with him two others, one on either side, with Jesus in the middle. *John* 19:1-5, 16-18

After this, aware that everything was now finished, in order that the scripture might be fulfilled, Jesus said, "I thirst." There was a vessel filled with common wine. So they put a sponge soaked in wine on a sprig of hyssop and put it up to his mouth. When Jesus had taken the wine, he said, "It is finished." And bowing his head, he handed over the spirit. Now since it was preparation day, in order that the bodies might not remain on the cross on the sabbath, for the sabbath day of that week was a solemn one, the Jews asked Pilate that their legs be broken and they be taken down. So the soldiers came and broke the legs of the first and then of the other one who was crucified with Jesus. But when they came to Jesus and saw that he was already dead, they did not break his legs, but one soldier thrust his lance into his side, and immediately blood and water flowed out... For this happened so that the scripture passage might be fulfilled: "Not a bone of it will be broken." And again another passage says: "They will look upon him whom they have pierced." *John* 19:28-34,36-7

The Tomb Narratives

When it was evening, there came a rich man from Arimathea named Joseph, who was himself a disciple of Jesus. He went to Pilate and asked for the body of Jesus; then Pilate ordered it to be handed over. Taking the body, Joseph wrapped it [in] clean linen and laid it in his new tomb that he had hewn in the rock. Then he rolled a huge stone across the entrance to the tomb and departed. *Matthew* 27: 57-60

Joseph of Arimathea, a distinguished member of the council, who was himself awaiting the kingdom of God, came and courageously went to Pilate and asked for the body of Jesus. Pilate was amazed that he was already dead. He summoned the centurion and asked him if Jesus had already died. And when he learned of it from the centurion, he gave the body to Joseph. Having bought a linen cloth, he took him down, wrapped him in the linen cloth and laid him in a tomb that had been hewn out of the rock. Then he rolled a stone against the entrance to the tomb. *Mark* 15:43-46

Now there was a virtuous and righteous man named Joseph who, though he was a member of the council, had not consented to their plan of action. He came from the Jewish town of Arimathea and was awaiting the kingdom of God. He went to Pilate and asked for the body of Jesus. After he had taken the body down, he wrapped it in a linen cloth and laid him in a rock-hewn tomb in which no one had yet been buried. It was the day of preparation, and the sabbath was about to begin. The women who had come from Galilee with him followed behind, and when they had seen the tomb and the way in which his body was laid in it, they returned and prepared spices and perfumed oils. Then they rested on the sabbath according to the commandment. *Luke* 23:50-56

After this, Joseph of Arimathea, secretly a disciple of Jesus for fear of the Jews, asked Pilate if he could remove the body of Jesus. And Pilate permitted it. So he came and took his body. Nicodemus, the one who had first come to him at night, also came bringing a mixture of myrrh and aloes weighing about one hundred pounds. They took the body of Jesus and bound it with burial cloths along with the spices, according to the Jewish burial custom. Now in the place where he had been crucified there was a garden, and in the garden a new tomb, in which no one had yet been buried. So they laid Jesus there because of the Jewish preparation day; for the tomb was close by. *John* 19:38-42

Then they returned from the tomb and announced all these things to the eleven and to all the others. The women were Mary Magdalene, Joanna, and Mary the mother of James; the others who accompanied them also told this to the apostles, but their story seemed like nonsense and they did not believe them. But Peter got up and ran to the tomb, bent down, and saw the burial cloths alone; then he went home amazed at what had happened. *Luke* 24:9-12

So Peter and the other disciple went out and came to the tomb. They both ran, but the other disciple ran faster than Peter and arrived at the tomb first; he bent down and saw the burial cloths there, but did not go in. When Simon Peter arrived after him, he went into the tomb and saw the burial cloths there, and the cloth that had covered his head, not with the burial cloths but rolled up in a separate place. *John* 20:3-7

The Shroud's Narrative

How does all this match up with the image of the Man of the Shroud? As doctors and scientists have testified, the wounds of the Man of the Shroud mirror the descriptions of the Passion in the Gospels. The accounts of the burial shroud in the Gospels, however, have no bearing on the Shroud itself beyond confirming the existence of a linen cloth wrapping the body of Jesus. Since the various Gospel shroud references come to us in different contexts, exegetes can assist us in understanding them.

We first consider the Shroud's Passion story. In recent years (1986 and 2006), two of the most prestigious medical journals in the world, the *Journal of the Royal Society of Medicine* in the UK and the *Journal of the American Medical Association* (JAMA), have published papers reconstructing Jesus' scourging, crowning with thorns, carrying of the cross and crucifixion. These and other medical reports will be treated in more detail in a subsequent chapter but at this point we will cite an overview of the Passion as it appears on the Shroud.

Archaeologist William Meacham summarizes what seems obvious at first glance in the journal *Current Anthropology*:

> Under the direction of Yves Delage, professor of comparative anatomy, a study was undertaken of the physiology and

pathology of the apparent body imprint and of the possible manner of its formation. The image was found to be anatomically flawless down to minor details: the characteristic features of rigor mortis, wounds, and blood flows provided conclusive evidence to the anatomists that the image was formed by direct or indirect contact with a corpse, not painted onto the cloth or scorched thereon by a hot statue (two of the current theories). On this point all medical opinion since the time of Delage has been unanimous (notably Hynek 1936; Vignon 1939; Moedder 1949; Caselli 1950; La Cava 1953; Sava 1957; Judica-Cordiglia 1961; Barbet 1963 ; Bucklin 1970; Willis, in Wilson 1978; Cameron 1978; Zugibe, in Murphy 1981)."...

The wounds seen in the Shroud image correspond perfectly with those of Christ recorded in the Gospel accounts: beating with fists and blow to the face with a club, flogging, "crown of thorns," nailing in hands (Aramaic *yad*, including wrists and base of forearm) and feet, lance thrust to the side (the right side, according to tradition) after death, issue of "blood and water" from the side wound, legs unbroken, McNair (1978:23) contends that such an exact concordance could hardly be coincidental: 'it seems to me otiose, if not ridiculous, to spend time arguing ... about the identity of the man represented in the Turin Shroud. Whether genuine or fake, the representation is obviously Jesus Christ.'

Expanding further, he notes:

Death had occurred several hours before the deposition of the corpse, which was laid out on half of the Shroud, the other half then being drawn over the head to cover the body. It is clear that the cloth was in contact with the body for at least a few hours, but not more than two to three days, assuming that decomposition was progressing at the normal rate. Both frontal and dorsal images have the marks of many small drops of a postmortem serous fluid exuded from the pores. There is, however, no evidence of initial decomposition of the body, no issue of fluids from the orifices, and no decline of rigor mortis leading to flattening of the back and blurred or double imprints.

Rigor mortis is seen in the stiffness of the extremities, the retraction of the thumbs (discussed below), and the distention of the feet. It has frozen an attitude of death while hanging by the arms; the rib cage is abnormally expanded, the large pectoral muscles are in an attitude of extreme inspiration (enlarged and drawn up toward the collarbone and arms), the lower abdomen is distended, and the epigastric hollow is drawn in sharply. The protrusion of the femoral quadriceps and hip muscles is consistent with slow death by hanging, during which the victim must raise his body by exertion of the legs in order to exhale.

The evidence of death in a position of suspension by the arms coupled with the characteristic wounds and blood flows indicate that the individual had been crucified. The rigor mortis position of outstretched arms would have had to be broken in order to cross the hands at the pelvis for burial, and a probable result is seen in the slight dislocation of the right elbow and shoulder. The feet indicate something of their original positioning on the cross, the left being placed on the instep of the right with a single nail impaling both. Apparently there was some flexion of the left knee to achieve this position, leaving the left foot somewhat higher than the right. ...

Of greatest interest and importance are the wounds. As with the general anatomy of the image, the wounds, blood flows, and the stains themselves appear to forensic pathologists flawless and unfakeable. 'Each of the different wounds acted in a characteristic fashion. Each bled in a manner which corresponded to the nature of the injury. The blood followed gravity in every instance' (Bucklin 1961:5). The bloodstains are perfect, bordered pictures of blood clots, with a concentration of red corpuscles around the edge of the clot and a tiny area of serum inside. Also discernible are a number of facial wounds, listed by Willis (cited in Wilson 1978:23) as swelling of both eyebrows, torn right eyelid, large swelling below right eye, swollen nose, bruise on right cheek, swelling in left cheek and left side of chin.

The body is peppered with marks of a severe flogging estimated at between 60 and 120 lashes of a whip with two or three studs at the thong end. Each contusion is about 3.7 cm long, and these are found on both sides of the body from the shoulders to the calves, with only the arms spared. Superimposed on the marks of flogging on the right shoulder and left scapular region are two broad excoriated areas, generally considered to have resulted from friction or pressure from a flat surface, as from carrying the crossbar or writhing on the cross. There are also contusions on both knees and cuts on the left kneecap, as from repeated falls. …

The pathology described thus far may well have characterized any number of crucifixion victims, since beating, scourging, carrying the crossbar, and nailing were common traits of a Roman execution. The lacerations about the upper head and the wound in the side are unusual and thus crucial in the identification of the Shroud figure. The exact nature of these wounds, especially whether they were inflicted on a living body and whether they could have been faked, is highly significant. Around the upper scalp and extending to its vertex are at least 30 blood flows from spike punctures. These wounds exhibit the same realism as those of the hand and feet: the bleeding is highly characteristic of scalp wounds with the retraction of torn vessels, the blood meets obstructions as it flows and pools on the forehead and hair, and there appears to be swelling around the points of laceration (though Bucklin [personal communication, 1982] doubts that swelling can be discerned). Several clots have the distinctive characteristics of either venous or arterial blood, as seen in the density, uniformity, or modality of coagulation (Rodante 1982). …

Between the fifth and sixth ribs on the right side is an oval puncture about 4.4 X 1.1 cm. Blood has flowed down from this wound and also onto the lower back, indicating a second outflow when the body was moved to a horizontal position. All authorities agree that this wound was inflicted after death, judging from the small quantity of blood issued, the separation of clot and serum, the lack of swelling, and the deeper color and more viscous consistency of the blood. Stains of a body fluid are intermingled with the blood, and numerous theories have been offered as to its origin: pericardial fluid (Judica, Barbet), fluid from the pleural sac (Moedder), or serous fluid from settled blood in the pleural cavity (Saval, Bucklin).

So convincing was the realism of these wounds and their association with the biblical accounts that Delage, an agnostic, declared them "a bundle of imposing probabilities" and concluded that the Shroud figure was indeed Christ. His assistant, Vignon (1937), declared the Shroud's identification to be 'as sure as a photograph or set of fingerprints.'[1]

Meacham concludes that "the exact correspondence of the wounds of Christ with those of the Shroud man is of supreme importance; if genuine, the Shroud would provide a most extraordinary archaeological reflection of the crucifixion accounts rendered by the evangelists."

Harold Attridge, dean of the Yale Divinity School and a well-known New Testament scholar, said of the Shroud: "It could well be the burial cloth of Jesus – I wouldn't discount that possibility. ... However this image was formed, it was formed in a way that's compatible with the ancient practice of Crucifixion. ... And the blood stains, for instance, are clearly not paint."[2]

The Burial Cloths

With regard to Jesus' burial, the first three Gospels clearly note that his body was wrapped in a linen cloth. The Gospel of John adds more detail to this description. Meacham points out that

> Greater difficulties are encountered in John's descriptions of the burial linens. The synoptic Gospels record that the body was wrapped or folded in a fine linen *sindon* or sheet. Although the traditional idea is that this sheet was wound around the body, there is no difficulty in reconciling it with the Shroud. John (20:5-8) describes the body as 'bound' with *othonia*, a word of uncertain meaning generally taken as 'cloth' or 'cloths.' In the empty tomb he relates seeing 'the *othonia* lying there, but the napkin (*soudarion*) which had been over the head not lying with the *othonia* but folded [or rolled up] in a place by itself.' To elucidate this passage, almost as many theories as there are possibilities have been put forward. One which would exclude the Shroud is that the linen sheet was cut up into bands to wrap around the corpse, but most exegetes reject this notion. The fact that Luke describes the body as wrapped in a *sindon* and then relates that the *othonia* were seen in the empty tomb is taken by some as an equation of the two, by others as a distinction. Most commentators identify the Shroud with the *sindon* and offer one of the following interpretations: (1) The *othonia* is the Shroud, the *soudarion* is a chin band tied around the head to hold up the lower jaw, and the hands and feet were bound with linen strips. In the account of Lazarus, a *soudarion* is mentioned 'around his face,' and his hands and feet are bound with *keiriai* (twisted rushes). Three-dimensional projections of the Shroud face have indicated a retraction of beard and hair where a chin band would have been tied. The Greek *soudarion* is clearly a kerchief or napkin. (2) The *soudarion* is the Shroud, and the *othonia* are bands used to tie up the body. In the vernacular Aramaic, *soudara* included larger cloths, and the phrases 'over his head' and 'rolled up in a

place by itself" suggest an item more substantial than a mere kerchief.³

Arnold Lemke notes that John and the other Gospels are talking of two different things given the distinction in terms between their burial accounts:

> Let's begin with the accounts of the three synoptic gospels where the terminology is simple and clear: Jesus' body was "wrapped" (ἐνειλέω, ἐντυλίσσω) in a "large linen cloth" (σινδών) and then placed into the grave. The Greek is quite straightforward here and provides a clear picture of what occurred. The problem comes with the parallel account from John (John 19:39-40). If John is here describing just the same actions and materials as the synoptics (which seems to be the presumption of most translators) then there would be an apparent conflict—a tying or wrapping up in "ὀθόνια" (strips of linen). But, looking more closely, we note that John actually uses quite different terminology from the others. He uses "δέω" instead of one of the three Greek words used specifically for wrapping; the basic meaning of "δέω" is to "tie, bind." He uses "ὀθόνια" and "κειρία"—"small linen strips," "cords"—to describe the material used for such a tying here and in the similar account of Lazarus (John 11:44). It is quite a stretch to try make these come out the same as "σινδών"—"large linen cloth"—as in the first three gospels.⁴

By comparing John's description of Jesus' burial to his account of Lazarus' rising, Lemke explains that John "was not in fact speaking to exactly the same part of the burial activity as the other three writers ... but to a quite different aspect of the process of which he personally was aware and which is not that well known today. ... [H]e was speaking to something quite different—a tying of the limbs to hold them in position at the time of burial due to rigor mortis rather than a separate wrapping or covering of the entire body with strips."

Maurus Green writes that:

"St Ephrem is the first writer we know of to identify *sindon* and *soudarion*. From the seventh century the Latin equivalent *sudarium* (and equivalents in all Romance languages, Georgian and Armenian) is used to translate both shroud and smaller face cloths, including Veronicas. In Syriac, Arabic and Aramaic, the vernacular of Palestine, equivalents of *sudarium* designated a square cloth used as a skirt, wide mantle, or ample veil over the head and enveloping the wearer. (Wuenschel cites Abbe Levesque's 'Le Suaire de Turin et L'Evangile', Nouvelle Revue Apologetique 1 (1939) 228.) The Abbé thinks that John's *soudarion* used in the burials of Lazarus and Christ should be interpreted in this Semitic sense, since the fourth Gospel abounds in Aramaisms. In support he refers to the current practice of the Druzes, ancient inhabitants of the Lebanon, who fold a shroud over the head down to the feet and tie it with bands at neck, feet and hand levels. He equates the bands with the *keiriai* of John 11:44, which kept Lazarus bound. He suggests that the *othonia* in the case of Christ would include the *keiriai* and the *soudarion* which, if used in the Semitic sense, would be the equivalent of the Synoptists' *sindon*."[5]

Gary Habermas and Kenneth Stevenson consider the use of *sindon* in the first three Gospels and the plural *othonia* in John to describe the burial cloths. They suggest that "*othonia* refers to *all* the grave cloths associated with Jesus' burial – the large *sindon* (the shroud), as well as the smaller strips of linen that bound the jaws, the hands and the feet. This interpretation of *othonia* is supported by Luke's use of the word. He says (23:53) that Jesus was wrapped in a *sindon*, but later (24:12) that Peter saw the *othonia* lying in the tomb after Jesus' resurrection. Luke, then, uses *othonia* as a plural term for all the grave cloths, including the *sindon*."[6]

Habermas and Stevenson further observe that "archaeological excavations at the Qumran community found that the Essenes buried their dead in the way represented on the Shroud. Several skeletons were found lying on their backs, faces pointing upward, elbows bent outward, and their hands covering the pelvic region. The protruding elbows rule out an Egyptian type mummified burial. Also very instructive is the Code of Jewish Law, which discusses burial procedures in its 'Law of Mourning.' It instructs that a person executed by the government was to be buried in a single sheet. This is another parallel with the Shroud."

Further, "Although the New Testament's description of typical first-century Jewish burial customs is not overly detailed, it does give the general features. The body was washed (Acts 9:37) and the hands and feet were bound (John 11:44). A cloth handkerchief (Greek, sudarion) was placed 'around' the face (John 11:44, 20:7). The body was then wrapped in clean linen, often mixed with spices (John 19:39-40), and laid in the tomb or grave."[7]

In sum, the Gospel narratives are embedded in and embodied by the Shroud. The Shroud image, in particular, uniquely testifies to the truth of the Gospel story. Uniquely because it is a image like no other before or since. More needs to be said about this.

2.2 Ecce Homo

Three properties of the Shroud image set it apart:

- It acts like a photographic negative but it was created centuries before the advent of photography
- It contains three-dimensional information: it is not just a "picture"
- It was formed from the microbiome of the Man of the Shroud.

We have seen how and why the image's microbiome-centered origin explains its photographic properties: the oxidation of the unsaturated fatty acids resulted in the formation of the image and variations in the amounts of these acids created lighter and darker areas. The same is true of its three-dimensional properties: this 3D property is a result of the varying densities of microbes accumulating in different parts of the body. In fact, as noted earlier, given their viscous polysaccharides, the skin bacteria once dried will form a rigid 3 D structure.

Informatics and multi-media expert Nello Balossino argues that the photographic and three dimensional properties of the Shroud image make it virtually impossible to replicate:

> In terms of formation [in normal photographs], an image is created through the interaction of light energy coming from a setting through the acquisition system. But it is only the light intensity that is recorded, not the phase in which the depth is codified: a photographic negative does not therefore possess three-dimensional information.

The image on the Shroud behaves like a negative ... [But] an ordinary photographic negative does not reproduce three-dimensional information. The image on the Shroud contains this information, which is codified in a series of nuances. In other words, what we have before us is an image formed through a three-dimensional process, which cannot yet be explained and simulated in practice in order to obtain replica images of the Shroud. The difference in tonality between the light and dark elements of the image is so low that the eye is only able to perceive the general features of a human face, but the details are not easy to make out. In fact the light distribution on the face depicted in the image is exactly the opposite to what we perceive in reality, with protruding elements appearing darker than the hollows of the face. The inversion process presents the face of a man from a real-life perspective.

I have examined the various attempts through the use of different techniques to reproduce the Turin Shroud. In none of these cases did the images last in time or contain the three-dimensional information contained in the Shroud and the features that are unique to the image such as the superficiality in the chromatic variation of the cloth fibres and their integrity. These details certainly make it harder to justify the explanation according to which the Shroud image is a medieval forgery.[1]

From the properties of the image, we turn now to its content. The Shroud actually has two kinds of images on it: blood stains and a body image. It is clear that the bloodstains came first because there is no body image under them.

Giulio Fanti, a professor at the University of Padua, observes that "The following facts have been verified: the bloodstains, transposed to the linen fabric by fibrinolysis, were impressed on the Shroud before the body image formed, since there is no body image under them; the processes of redissolving and transposition of blood in a damp environment may occur after a period of 10 hours; the body of the

Man remained in the Shroud for less than 40 hours, because no signs of putrefaction can be found."[2]

In studying the bloodstains and the body image, we should bear in mind that the appearance of the body that is visible on the Shroud will be noticeably affected by numerous factors ranging from its manner of death to the stiffening of rigor mortis to actions taken before burial. Distortions there will be given the setting, the medium and the state of the body.

Too often, Internet flame-throwers rush in with split-second judgments where medical experts fear to tread. Evidence and expertise count for nothing in discussions characterized by flippancy, invective and mindless monologues. Our approach here is different: we listen to the specialists when we consider specialized areas; we seek and study evidence available from a multitude of relevant sources; we engage serious skeptics in those domains where their challenge is most powerful; we differentiate between facts and speculation; and we always strive to see things in terms of the big picture.

Body Image and Bloodstains

Based on their research, a team of Italian scientists which included the prolific Fanti concluded that the Shroud image

> "indicates a perfect immobility of the human body wrapped in it: the very detailed image would be impossible otherwise. The facial image shows the typical [face] of a corpse. ... The body shows an extreme rigidity, producing an S shape with the head bent towards the chest, and the chest towards the abdomen. The chest is shown in an intense inhaling contraction, the abdomen protruding outwards, and the legs half bent, still retaining the asymmetry caused by their position on the cross. There are bloodstains due to human blood, from wounds inflicted upon a living person and also post mortem flows. The side wound is certainly post mortem: the edges of the wound are clearly undulated, since they did not retract as would be the case if the subject were alive when the blow was inflicted. Blood clots

and serum are clearly separated, so that compared with blood seen on the arms they seem lacking in structure. This is true for the stains in the lower part of the wound, and in the flow that crosses the back, that occurred about two hours after death.[3]

An analysis of the body of the Man of the Shroud must recognize first that it had been washed before burial. The evidence for this arises from three sources:

- ➢ Jewish burial practices,
- ➢ the confirmation from the Gospel of John and similar texts and
- ➢ the scientific study of the image itself.

Jewish burial customs, as laid out for instance in the second century A.D. Mishna, stipulate that even on the Sabbath "one may anoint him [the dead person] with oil and wash him." In addition, "one may tie up the jaw, not in order that it should close but that it should not further open."[4]

John 19:40 specifically states that the burial of Jesus was carried out "according to the Jewish burial custom." The apocryphal second century Gospel of Peter, known to have some historical elements, says about the burial that Joseph "took the Lord and washed him and wrapped him in linen and brought him unto his own sepulcher."[5] The Stone of Unction/Anointing at the entrance of the Church of the Holy Sepulcher commemorates the original stone on which Jesus' body was washed and anointed prior to burial. Apart from religious reasons, washing was essential given that dirt and debris would have attached itself to Jesus' body during his way of the cross and after.

The scientific evidence for the washing of the body has been well presented by Dr. Frederick Zugibe who we will cite on various issues. Zugibe was a forensic pathologist and cardiologist who was Chief Medical Officer of Rockwood County, New York (1969 to 2003), Adjunct Associate Professor at the Columbia University College of Physicians and Surgeons and Director of Cardiovascular Research with the US Veterans Administration. He was considered one of the leading forensic pathologists in the US and directed some 10,000 autopsies. In discussing the washing, Zugibe writes:

> Imprints depicting the various wounds that had been inflicted on the Man of the Shroud include numerous dumbbell-shaped scourge marks over the trunk, an exact pattern of rivulets of blood on the left arm, a single tortuous flow of blood on the forehead, a precise bifurcation pattern on the back of the hand and a small clump of blood on the heel. Studies of these patterns with ultraviolet light are even more vivid in terms of preciseness; the scourge marks show well defined borders and fine scratch-like markings appear to be mingled in-between. ...
>
> If the deceased individual had not been washed, these well-defined wound patterns depicted on the Shroud could not be present. First of all, most of the blood within the scourge wounds of the victim would have been clotted and the blood located both at the periphery and outside of the wounds would have dried long before the victim was placed on the cross. ...
>
> Forgetting all of the other wounds, no one would argue that the scourge wounds were made and clotting begun several hours prior to death. Moreover, most forensic experts agree that the Man of the Shroud shows evidence of rigor mortis because of the bent knees and absence of a neck, therefore indicating that the crucified was dead for some time before being taken down from the cross. Thus, according to the studies of Lavoie's group, these *perfectly defined wounds* should not have transferred at all. Yet many of the scourge wounds on the Shroud of Turin are extremely distinct corresponding to dumbbell shaped wounds. ... However, if the body was washed, the dried blood around the wounds would be removed causing an oozing of bloody material within the wounds resulting in the production of relatively good impressions of the wound.

The expertise concerning blood flow patterns is in the area of forensic pathology. The forensic pathologist is frequently called in to court to provide expert testimony regarding blood flow patterns and wound characteristics and to render an opinion regarding the mechanism, manner and cause of death, concerning these circumstances. This applies to the Man of the Shroud who was apparently scourged, crowned with thorns, nailed through the hands and feet with large square nails and suspended by the hands for several hours.

A forensic evaluation of the crucifixion reveals that every movement during the entire time the crucified was on the cross would have restarted bleeding in the hand and foot wounds. The body unquestionably would have been literally covered with blood because the heart pumps about 4500 gallons of blood through more than 60,000 miles of large and small blood vessels throughout the whole body each day. Instead of the very exact imprints of the wounds, the Shroud would instead bear large indistinct masses of blood over the entire image including the face, arms, hands, feet and trunk. Every practicing forensic pathologist knows that even tiny wounds may bleed profusely during heart activity and observes the end results of bleeding from wounds of practically every type on a daily basis. ...

It is also of importance to note that scourge markings were made many hours prior to removal from the cross so that encrusted clots would have formed in the wounds therefore making it difficult to understand how the scourge marks would have left such precise imprints. Every forensic pathologist that I consulted with, agreed that the wounds would have caused a large amount of bleeding, and the body had to be washed to account for the preciseness of the wounds. In the December 1980 issue of *Medical World News*, Dr. Michael Baden, a forensic pathologist and the former Chief Medical Examiner of New York City agreed that if the Shroud is genuine, the body must have been washed. He also added that if the body was washed there might be some oozing from the wounds.

Jewish law forbade washing away the life-blood of a corpse but this refers to any blood that flows after death and not before. Zugibe writes that

> "the only blood flow after death included a small amount of a watery, blood tinged discharge that extruded from the lance wound. The amount of watery effusate and blood exuding from the lance wound had to be small because immediately upon the introduction of the spear into the chest cavity, the lungs would collapse (pneumothorax) due to the increased atmospheric pressure thereby causing the fluid level to immediately drop because there would *now be* more space in the chest cavity. Therefore, the only fluid extruding out of the wound would be due to the initial penetration by the lance (immediately prior to collapse) and the small amount of blood from the right atrium of the heart contained on the spear tip by a quick, jerking motion following the sudden thrust. In my opinion, this amount is less than *the quarter log* quantitated as the minimum amount of blood after death that is required to become unclean according to the Mishna and Talmud. All of the other blood on the body prior to washing would have been present prior to death and could be washed according to the Jewish prescription. If a rapid burial ceremony was necessary, the washing ritual could be effected in a few minutes. The body could then be washed in a few minutes.
>
> Even if more than a *quarter log* extruded from the lance wound, the small amount of blood around the spear wound could easily be avoided during the washing procedure. The act of washing would then cause an oozing from each of the [other] wounds thereby accounting for the imprints at their locations consistent with those on the Shroud. The blood that oozed from these wounds could not be subsequently washed

because this is considered *unclean* blood. This would therefore conform to the requirements of Jewish Law, account for the well defined wounds depicted on the Shroud of Turin, and provide an explanation that would satisfy the practicing forensic pathologist.[6]

Biblical scholar Raymond Brown emphasizes that there are constraints on what we know about first century Jewish burial practices:

We must be careful to recognize limitations in our knowledge of burial practices in Jesus' lifetime. Even before recent sensitivity about the limited applicability of the Mishna to Jesus' time, and therefore about mishnaic rules for burying the bodies of the condemned, Buckler recognized that the references to burial in Josephus indicated a different situation in the 1st cent. from that envisioned by later information.[7] ... Some aspects of the mishnaic practice were surely ideal or reflect a post-NT situation.[8]

Thomas de Wesselow argues that the nature of the body and blood images testifies to the authenticity of the Shroud. The blood wounds, for instance, could not have been painted by a medieval artist for various reasons:

- The nail wounds seen on the Shroud enter through the wrists but ALL medieval paintings depict these in the middle of the palms.
- The asymmetrical patterns of blood flows on the Shroud would not be found in a medieval painting.
- The lack of clarity in the nail wound in the feet again make it unlikely that this originated in the Middle Ages.
- The "impressionistic effect of the Shroud might seem 'convincing' to us today, but it would have been nonsensical and shocking – even unthinkable – in the Middle Ages, when artists were bound by theological and devotional requirements."[9]

- The scourge wounds are very visible in the Shroud but the overwhelming majority of medieval paintings of the dead or dying Jesus show no scourge wounds.
- The Shroud accurately shows that the victim bore only the horizontal beam of the cross whereas medieval art invariably showed him carrying the entire cross.
- There are numerous small puncture wounds on the head of the Man of the Shroud that indicate a cap of thorns covering his head and not the kind of "crown of thorns" found in medieval art.

The other obvious and often mentioned feature is the lack of clothing on the Man of the Shroud – this too is unlike anything in medieval art.

De Wesselow says "the overwhelming majority of those who consider the matter carefully (including atheists, agnostics and non-Catholic Christians with a healthy disregard for religious relics) conclude that the Shroud might very well be what it purports to be: the winding sheet of Jesus. And the primary evidence that leads to this conclusion is the pattern of injuries apparent on the cloth. *Far from being too good to be true, the Shroud's blood-image seems too good to be false.* [Emphasis added.] Medically attested as a convincing representation of severe injuries and chemically proven to consist of blood, there is no rational reason to deny that the bloodstains are natural traces of a man crucified in accordance with Roman practice, crowned with thorns and buried as a Jew. The notion that such a physiologically and archeologically accurate image would have been painted (in blood) by a medieval artist is patently absurd. As the great Jewish art historian Ernst Kitzinger is reported to have said, 'there are no paintings that have blood marks like those of the shroud. You are free to look as you please but you won't find any.' If the blood-image is not painted, it must derive from a genuine death and burial."[10]

Unfazed by the evidence of the blood-image, skeptics have claimed that the body-image is a medieval painting. But no major art historian has endorsed this view. De Wesselow remarks, "From an art-historical point of view, the idea that the Shroud's body-image was painted shortly before 1356, the approximate date of its first display in Lirey, is untenable. The Shroud's image is quite unlike any painting of the period – or, indeed, of any period. In the words of Ernst Kitzinger, 'The Shroud of Turin is unique in art. It doesn't fall into any artistic category.'"[11]

The image as it appears to the naked eye is a vague blur of a human body. It is only when photographed that it is seen to be the striking image with which we are familiar. The normally visible image, says de Wesselow, "is not the sort of image with which any medieval artist would have been concerned. It is not the sort of image that a medieval artist would have produced, and the fact that it has an invisible structure, which could not have been appreciated then by anyone, is utterly damning for the painting hypothesis. ... How could a fourteenth-century artist have achieved such a transformation, even if he had wanted to?"

Beyond the artistic issues, there are scientific objections as well. "For a start, there is no sign of any pigment or binding medium. No artist could or would have colored just the topmost fibers of the cloth with an undetectable, non-liquid substance and then turned it over and done the same in the region of the hair. STURP also observed that the colored fibers remained the same next to the 1532 scorches, indicating they had nothing to do with an organic pigment, which would have been discolored by the heat. The body image was equally unaffected by the water damage represented by the diamond-shaped water stains meaning the color is insoluble, ruling out the use of a water-based paint medium. Finally, ... the coloring of the body-image is entirely directionless, i.e. there is no sign of any brushwork."[12]

Rigor Mortis

As we saw, Meacham points out that the image on the Shroud shows "the characteristic features of rigor mortis."[13]

The characteristics of the body image show the effect of rigor mortis:

- The positioning of the feet indicate that "the man wrapped in the Shroud would seem to have died with one foot crossed over the other, his left leg fractionally bent, a position maintained after death. ... Doctors who have studied the Shroud are in agreement that rigor mortis is implied by the imprints of the legs and that the pose implies crucifixion."[14]

- The positioning of the hands over the groin suggest that the hands stretched out in rigor mortis after the crucifixion were forcibly re-positioned.
- There is no neck visible on the man of the Shroud because during crucifixion "his head would have dropped forward at an angle of about forty degrees to the vertical, bringing his chin down close to the base of his throat. The head remained bowed as the neck muscles developed rigor and was then fixed in that position even after the body was taken down from the cross and laid horizontally."[15]

Like Zugibe, Robert Bucklin, another Medical Examiner, points to the effect of rigor mortis:

The body structure is anatomically normal, representing a well-developed and well-nourished individual with clearly identifiable head, trunk, and extremities. The body appears to be in a state of rigor mortis which is evidenced by an overall stiffness as well as specific alterations in the appearance of the lower extremities from the posterior aspect. The imprint of the right calf is much more distinct than that of the left indicating that at the time of death the left leg was rotated in such a way that the sole of the left foot rested on the ventral surface of the right foot with resultant slight flexion of the left knee. That position was maintained after rigor mortis had developed.[16]

In their "computerized anthropometric analysis" of the Man of the Shroud, three Italian scientists refer first to the work done by a pioneer in this kind of study. "Giulio Ricci carried out a very detailed analysis, but with some limitations ... he supposed that the sheet was softly placed on the body outline and that the body had not been laid out flat, both because of the position taken on the cross and because of rigor mortis; moreover, thanks to the blood traces and trickles, he was able to deduce the presence of folds on the sheet and the form this took on the body. Therefore the measurements realized directly on it and not corrected constitute only the linear development of the body outline." In their own analysis they take into account other effects such as the "effect of the forward bending of the head" and "probable "S" bending of the backbone."[17]

The microbial factors behind the formation of the image must also be considered. With microbes, environmental conditions can certainly cause distortions. All that these microbes (*Staphylococcus epidermidis*) need to move on the linen is water, diffused nutrients from the body, and room temperature as shown in the experiments in the first section. The experiments show that the bacteria can spread which means distortions are quite possible. This also explains how the image can form over a body part that does not make contact with the skin.

After reviewing the effects of rigor mortis and the impact of the positioning of the cloth, de Wesselow concludes that

> Every part of the body-image, then, even the remarkable image of the head, can be explained in terms of a stain produced by a real human body. Indeed, certain distortions and gaps in the image demand to be explained in these terms. The missing feet and neck of the frontal figure, the blank areas around the hands and forearms, the soles of the feet in the dorsal figure: these speak of the cloth having been draped loosely over the body of a man, not tinted by a medieval artist or subjected to a primitive form of photography.
>
> Morever, we can tell that the body was fixed in rigor mortis and that, when the man died, he was suspended vertically, his feet crossed one over the other, his head bowed. There can be little doubt he died of crucifixion. It may be regarded as certain, therefore, that the body-image was imprinted on the Shroud by the same corpse as was responsible for the blood-image. The two super-imposed images are in perfect accord: they reinforce each other, proving beyond doubt that the Shroud really did enfold the dead body of a crucified man."[18]

2.3 Doctors' Diagnoses

Since the Man of the Shroud was subjected to lesion and laceration, trauma and shock, he suffered from internal and external injuries that could be medically diagnosed. Such a diagnosis would make sense of what we see on the Shroud and the event it embedded. There is nothing religious about this. It is simply the practice of medicine. And, in fact, two of the world's leading medical journals have studied the suffering and death of Jesus with one specifically doing so in the context of the Shroud. Further, numerous medical professionals have given their considered analyses of the wounds and blood flows visible on the Shroud. In brief, we have a treasure trove of doctors' diagnoses when it comes to the Man of the Shroud. We have cited some of these diagnoses as they relate to the blood and body images but it would be helpful to have a more comprehensive analysis.

There is a reason for emphazing that these are doctors' diagnoses. Today's search engines and data repositories furnish Internet users with access to vast oceans of information once restricted to academia. Nevertheless, there are certain spheres where reliable judgments depend on actual hands-on experience. Medical diagnosis is one such. Only a lifetime of clinical experience – from feeling a pulse to treating a wound to extracting a tumor – can equip a person to diagnose or treat different medical conditions. Freedom of speech is obviously a right but it is no guarantee of truth or wisdom. And since we have to make efficient use of the time available for exploring urgent questions, we have no obligation to give "equal time" to raconteurs holding forth on specialized domains in which they have no previous experience (e.g., on a question such as the "look" of the wounds on the Shroud). Of course, it hardly needs to be said that specialists sometimes differ on key topics. In such instances, we have good reason to examine their differences.

What we are highlighting is the criticality of relying only on sources that have earned the right to be heard. There are numerous flat earth societies today but no serious astronomers who belong to them. We have no intellectual obligation to listen to these societies' arguments for a flat earth. Likewise, the only medical diagnoses of the Shroud image considered here are from medical professionals.

Appropriately enough, we will start with the analysis of the Shroud's Passion narrative in one of the world's leading medical journals, the *Journal of the American Medical Association*. The JAMA study, titled "On the Physical Death of Jesus Christ,"[1] was led by Mayo Clinic professor of Laboratory Medicine and Pathology William D. Edwards, the author of over 400 peer-reviewed publications. Below are excerpts:

> The Shroud of Turin is considered by many to represent the actual burial cloth of Jesus, and several publications concerning the medical aspects of his death draw conclusions from this assumption. The Shroud of Turin and recent archaeological findings provide valuable information concerning Roman crucifixion practices. The interpretations of modern writers, based on a knowledge of science and medicine not available in the first century, may offer additional insight concerning the possible mechanisms of Jesus' death.[2] ...
>
> At nearby Gethsemane, Jesus, apparently knowing that the time of his death was near, suffered great mental anguish, and, as described by the physician Luke, his sweat became like blood. Although this is a very rare phenomenon, bloody sweat (hematidrosis or hemohidrosis) may occur in highly emotional states or in persons with bleeding disorders. As a result of hemorrhage into the sweat glands, the skin becomes fragile and tender. Luke's description supports the diagnosis of hematidrosis rather than eccrine chromidrosis (brown or yellow-green sweat) or stigmatization (blood oozing from the palms or elsewhere).[3]

Flogging was a legal preliminary to every Roman execution. ... As the Roman soldiers repeatedly struck the victim's back with full force, the iron balls would cause deep contusions, and the leather thongs and sheep bones would cut into the skin and subcutaneous tissues. ... Pain and blood loss generally set the stage for circulatory shock. The extent of blood loss may well have determined how long the victim would survive on the cross.[4]

The severe scourging, with its intense pain and appreciable blood loss, most probably left Jesus in a preshock state. Moreover, hematidrosis had rendered his skin particularly tender. The physical and mental abuse meted out by the Jews and the Romans, as well as the lack of food, water, and sleep, also contributed to his generally weakened state. Therefore, even before the actual crucifixion, Jesus' physical condition was at least serious and possibly critical.[5]

Although the Romans did not invent crucifixions they perfected it as a form of torture and capital punishment that was designed to produce a slow death with maximum pain and suffering. ... It was customary for the condemned man to carry his own cross from the flogging post to the site of crucifixion outside the city walls. He was usually naked, unless this was prohibited by local customs. Since the weight of the entire cross was probably well over 300 lb (136 kg), only the crossbar was carried. The patibulum, weighing 75 to 125 lb. (34 to 57 kg), was placed across the nape of the victim's neck and balanced along both shoulders. Usually, the outstretched arms then were tied to the crossbar.[6]

At the site of execution, by law, the victim was given a bitter drink of wine mixed with myrrh (gall) as a mild analgesic. The criminal was then thrown to the ground on his back, with his arms outstretched along the patibulum. The hands could be nailed or tied to the crossbar, but nailing apparently was preferred by the Romans. The archaeological remains of a crucified body, found in an ossuary near Jerusalem and dating from the time of Christ, indicate that the nails were tapered iron spikes approximately 5 to 7 in (13 to 18 cm) long with a square shaft 3/8 in (1 cm) across.[7]

Although the feet could be fixed to the sides of the stipes or to a wooden footrest (suppedaneum), they usually were nailed directly to the front of the stipes. To accomplish this, flexion of the knees may have been quite prominent, and the bent legs may have been rotated laterally.[8]

The length of survival generally ranged from three or four hours to three or four days and appears to have been inversely related to the severity of the scourging. However, even if the scourging had been relatively mild, the Roman soldiers could hasten death by breaking the legs below the knees (crurifragium or skelokopia).[9]

By Roman law, the family of the condemned could take the body for burial, after obtaining permission from the Roman judge. Since no one was intended to survive crucifixion, the body was not released to the family until the soldiers were sure that the victim was dead. By custom, one of the Roman guards would pierce the body with a sword or lance. Traditionally, this had been considered a spear wound to the heart through the right side of the chest—a fatal wound probably taught to most Roman soldiers. The Shroud of Turin documents this form of injury.[10]

The scourging prior to crucifixion served to weaken the condemned man and, if blood loss was considerable, to produce orthostatic hypotension and even hypovolemic shock. When the victim was thrown to the ground on his back, in preparation for transfixion of the hands, his scourging wounds most likely would become torn open again and contaminated with dirt. Furthermore, with each respiration, the painful scourging wounds would be scraped against the rough wood of the stipes. As a result, blood loss from the back probably would continue throughout the crucifixion ordeal.[11]

The actual cause of death by crucifixion was multifactorial and varied somewhat with each case, but the two most prominent causes probably were hypovolemic shock and exhaustion asphyxia. Other possible contributing factors included dehydration, stress-induced arrhythmias, and congestive heart failure with the rapid accumulation of pericardial and perhaps pleural effusions. Crucifracture (breaking the legs below the knees), if performed, led to an asphyxic death within minutes. Death by crucifixion was, in every sense of the word, excruciating (Latin, excruciatus, or "out of the cross").[12]

The soldiers broke the legs of the two thieves, but when they came to Jesus and saw that he was already dead, they did not break his legs. Rather, one of the soldiers pierced his side, probably with an infantry spear, and produced a sudden flow of blood and water. ... Although the side of the wound was not designated by John, it traditionally has been depicted on the right side. Supporting this tradition is the fact that a large flow of blood would be more likely with a perforation of the distended and thin-walled right atrium or ventricle than the thickwalled and contracted left ventricle.[13]

Some of the skepticism in accepting John's description has arisen from the difficulty in explaining, with medical accuracy, the flow of both blood and water. Part of this difficulty has been based on the assumption that the blood appeared first, then the water. However, in the ancient Greek, the order of words generally denoted prominence and not necessarily a time sequence. Therefore, it seems likely that John was emphasizing the prominence of blood rather than its appearance preceding the water. Therefore, the water probably represented serous pleural and pericardial fluid, and would have preceded the flow of blood and been smaller in volume than the blood. Perhaps in the setting of hypovolemia and impending acute heart failure, pleural and pericardial effusions may have developed and would have added to the volume of apparent water. The blood, in contrast, may have originated from the right atrium or the right ventricle or perhaps from a hemopericardium.[14]

Jesus' death after only three hours on the cross surprised even Pontius Pilate. The fact that Jesus cried out in a loud voice and then bowed his head and died suggests the possibility of a catastrophic terminal event.[15]

Jesus' death may have been hastened simply by his state of exhaustion and by the severity of the scourging, with its resultant blood loss and preshock state. The fact that he could not carry his patibulum supports this interpretation. The actual cause of Jesus' death, like that of other crucified victims, may have been multifactorial and related primarily to hypovolemic shock, exhaustion asphyxia, and perhaps acute heart failure. A fatal cardiac arrhythmia may have accounted for the apparent catastrophic terminal event.

The important feature may be not how he died but rather whether he died. Clearly, the weight of historical and medical evidence indicates that Jesus was dead before the wound to his side was inflicted and supports the traditional view that the spear, thrust between his right ribs, probably perforated not only the right lung but also the pericardium and heart and thereby ensured his death. Accordingly, interpretations based on the assumption that Jesus did not die on the cross appear to be at odds with modern medical knowledge.[16]

Another relevant paper, titled "Medical theories on the cause of death in crucifixion," appeared in the *Journal of the Royal Society of Medicine*. The paper is concerned with various theories of the cause of death in crucifixion and notes that "The postulated causes of death include cardiovascular, respiratory, metabolic, and psychological pathology." It observes that "The most detailed accounts of any one particular crucifixion are the biblical passages covering the death of Jesus of Nazareth; but we should not assume that this was by any means representative of all crucifixions." Although researchers have tried to re-enact the crucifixion up to a point, there are constraints on this kind of reconstruction: "The fact that none of the re-enactment research has actually crucified people means that these studies have only limited relevance to genuine cases. The absence of whipping, carrying a heavy cross, being nailed to it, the dehydration from water deprivation and hot sun, and the anxiety of their imminent death might all have resulted in somewhat different findings in the modern groups and crucifixion victims 2000 years ago."[17] Crucifixion came to an end in the Roman Empire after it was banned by the Emperor Constantine in the fourth century.

The classic medical work on the Shroud and the crucifixion is *A Doctor at Calvary* by the First World War battlefield surgeon Pierre Barbet. This moving book has had a transformational impact on thousands of readers since it first appeared in 1950. Numerous other medical professionals have written on the Shroud. The two most significant are medical examiners and forensic pathologists, Robert Bucklin of Los Angeles who personally performed 25,000 autopsies and Frederick Zugibe who, as we said, directed some 10,000 autopsies. Both were convinced of the authenticity of the Shroud and delved into the physiological details of the Passion as depicted in the Gospels and the Shroud. (It should be noted that Zugibe disagreed with Barbet on two issues: on the location of the nails in the hands and the cause of death at crucifixion.) Their diagnoses are the ones most worthy of attention given that they specialize in assessing the injuries and cause of death of murder victims – and here we are dealing with the most famous murder victim in history.

This is Bucklin's autopsy of the the Man of the Shroud:

For over 50 years as a Forensic Pathologist, I have been actively involved with the investigation of deaths which come under the jurisdiction of a coroner Medical Examiner. During that time, I have personally examined over 25,000 bodies by autopsy to determine the cause and manner of death.

For most of that same period of time, I have had an abiding interest in the study of the Shroud of Turin from a medical view point. ... The full body imprint, front and back, together with the individual characteristics of blood stains on the cloth, which represent specific types of injury, make it quite feasible for an experienced forensic pathologist to approach the examination of the Shroud image as would a medical examiner performing an autopsy on a person who has died under unnatural circumstances. ...

The first step in such an examination is to document physical features of the victim as accurately as possible. In the case of the image on the Shroud, it can be stated that the deceased person is an adult male measuring 71 inches from crown to heel and weighing an estimated 175 pounds. The body structure is anatomically normal, representing a well-developed and well-nourished individual with clearly identifiable head, trunk, and extremities. The body appears to be in a state of rigor mortis which is evidenced by an overall stiffness as well as specific alterations in the appearance of the lower extremities from the posterior aspect. The imprint of the right calf is much more distinct than that of the left indicating that at the time of death the left leg was rotated in such a way that the sole of the left foot rested on the ventral surface of the right foot with resultant slight flexion of the left knee. That position was maintained after rigor mortis had developed.

After an overall inspection and description of the body image, the pathologist continues his examination in a sequential fashion beginning with the head and progressing to the feet. He will note that the deceased had long hair, which on the posterior image appears to be fashioned into a pigtail or braid type configuration. There also is a short beard which is forked in the middle. In the frontal view, a ring of puncture tracks is noted to involve the scalp. One of these has the configuration of a letter "3". Blood has issued from these punctures into the hair and onto the skin of the forehead. The dorsal view shows that the puncture wounds extend around the occipital portion of the scalp in the manner of a crown. The direction of the blood flow, both anterior and posterior, is downward. In the midline of the forehead is a square imprint giving the appearance of an object resting on the skin. There is a distinct abrasion at the tip of the nose and the right cheek is distinctly swollen as compared with the left cheek. Both eyes appear to be closed, but on very close inspection, rounded foreign objects can be noted on the imprint in the area of the right and left eyes.

Upon examining the chest, the pathologist notes a large blood stain over the right pectoral area. Close examination shows a variance in intensity of the stain consistent with the presence of two types of fluid, one comprised of blood, and the other resembling water. There is distinct evidence of a gravitational effect on this stain with the blood flowing downward and without spatter of other evidence of the projectile activity which would be expected from blood issuing from a functional arterial source. This wound has all the characteristics of a postmortem type flow of blood from a body cavity or from an organ such as the heart. At the upper plane of the wound is an ovoid skin defect which is characteristic of a penetrating track produced by a sharp puncturing instrument.

There seems to be an increase in the anteroposterior diameter of the chest due to bilateral expansion.

The abdomen is flat, and the right and left arms are crossed over the mid and lower abdomen. The genitalia cannot be identified.

By examination of the arms, forearms, wrists, and hands, the pathologist notes that the left hand overlies the right wrist. On the left wrist area is a distinct puncture-type injury which has two projecting rivulets derived from a central source and separated by about a 10 degree angle. As it appears in the image, the rivulets extend in a horizontal direction. The pathologist realizes that this blood flow could not have happened with the arms in the position as he sees them during his examination, and he must reconstruct the position of the arms in such a way as to place them where they would have to be to account for gravity in the direction of the blood flow. His calculations to that effect would indicate that the arms would have to be outstretched upward at about a 65 degree angle with the horizontal. The pathologist observes that there are blood flows which extend in a direction from wrists toward elbows on the right and left forearms. These flows can be readily accounted for by the position of the arms which he has just determined.

As he examines the fingers, he notes that both the right and left hands have left imprints of only four fingers. The thumbs are not clearly obvious. This would suggest to the pathologist that there has been some damage to a nerve which would result in flexion of the thumb toward the palm.

As he examines the lower extremities, the medical examiner derives most of his information from the posterior imprint of the body. He notes that there is a reasonably clear outline of the right foot made by the sole of that foot having been covered with blood and leaving an imprint which reflects the heel as well as the toes. The left foot imprint is less clear and it is also noticeable that the left calf imprint is unclear. This supports the opinion that the left leg had been rotated and crossed over the right instep in such a way that an incomplete foot print was formed. In the center of the right foot imprint, a definite punctate defect can be noted. This puncture is consistent with an object having penetrated the structures of the feet, and from the position of the feet the conclusion would be reasonable that the same object penetrated both feet after the left foot had been placed over the right.

As the back image is examined, it becomes quite clear that there is a series of traumatic injuries which extend from the shoulder areas to the lower portion of the back, the buttocks, and the backs of the calves. These images are bifid [divided in two] and appear to have been made by some type of object applied as a whip, leaving dumbbell-shaped imprints in the skin from which blood has issued. The direction of the injuries is from lateral toward medial and downward suggesting that the whip was applied by someone standing behind the individual.

An interesting finding is noted over the shoulder blade area on the right and left sides. This consists of an abrasion or denuding of the skin surfaces, consistent with a heavy object, like a beam. Resting over the shoulder blades and producing a rubbing effect on the skin surfaces.

With this information available to him, the forensic pathologist can come to a reasonable conclusion as to the circumstances of death, including the posture of the deceased at the time the injuries were incurred. Chronologically, the whip like injuries to the back would have occurred earlier than other injuries which the pathologist has found. The individual would have been upright and with his arms above his head at the time the whipping occurred since no whip marks are found on the upper extremities.

The position of the puncture defects in the wrist, coupled with the blood flow towards the elbows, and also associated with the punctures of the feet, permit the pathologist to conclude that the victim was in an upright position with his arms extended when the blood flow took place. A crucifixion type posture would be the most plausible explanation for these findings.

The wound in the right side, since it is comprised of both blood and non-blood components, suggests to the forensic pathologist that the puncturing instrument released a watery type fluid from the body cavities as well as blood from the heart area.

One potential consideration would be that there was fluid in the chest cavity which was released by the penetrating instrument and this was followed by blood issuing from an area as the result of the heart being perforated.

At this point, the pathologist has garnered much information about the injuries to the body from a purely objective point of view. As a knowledgeable and expertly trained forensic pathologist he has the right and obligation to rely upon available historical and other evidentiary information in order to support or deny his impressions. He will avail himself of other scientific testing, including radiological studies and hematological and chemical testing of the substances which he has found on the body. By these tests, he will be able to confirm the presence of blood. He may also make other observations based on microscopic and genetic studies.

It is the ultimate responsibility of the medical examiner to confirm by whatever means are available to him the identity of the deceased, as well as to determine the manner of this death. In the case of Man on the Shroud, the forensic pathologist will have information relative to the circumstances of death by crucifixion which he can support by his anatomic findings. He will be aware that the individual whose image is depicted on the cloth has undergone puncture injuries to his wrists and feet, puncture injuries to his head, multiple traumatic whip-like injuries to his back and postmortem puncture injury to his chest area which has released both blood and a water type of fluid. From [these] data, it is not an unreasonable conclusion for the forensic pathologist to determine that only one person historically has undergone this sequence of events. That person is Jesus Christ.

As far as the mechanism of death is concerned, a detailed study of the Shroud imprint and the blood stains, coupled with a basic understanding of the physical and physiological changes in the body that take place during crucifixion, suggests strongly that the decedent had undergone postural asphyxia as the result of his position during the crucifixion episode. There is also evidence of severe blood loss from the skin wounds as well as fluid accumulation in the chest cavities related to terminal cardio-respiratory failure.[18]

We turn now to Zugibe's account of the Passion:

THE SCOURGING (flagellatio) was a brutal episode. The effects of the scourging appear very vivid on the Shroud showing dumbbell-type injuries, obviously caused by the flagrum which contains leather thongs with bits of metal or bone at the ends. The *crucarius* was tied by the hands to a fixed object like a pillar, bent over and lashed. The weight of the metal or bony objects would also carry them to the front of the body as well as the back and arms. The brutality of scourging can not be overestimated because these objects would penetrate the skin creating small lacerations (tears), contusions or welts. It is interesting that there are over a hundred lashes counted on the Shroud. Does this estimate conflict with the Deuteronomy dictate (25:3) not to exceed 40 lashes? The answer is simple. The flagrum consists of at least three thongs, each lash would cause three lash marks and 40 lashes times 3 would equal 120. These markings on the Shroud would be neither evidence of a bruise or welt as contended by some but instead they appear to be impressions of small breaks in the skin resulting in "patterned injuries" like we regularly see in the practice of forensic pathology as different instruments cause different patterns. These patterns on the Shroud are a result of impressions made by the blood present within the breaks in the skin. Such injuries are only seen at autopsy after gently washing the wounds otherwise there would be blood all over the body from these wounds obscuring the patterned impressions. When the body is initially washed, a fine oozing of blood within the wounds would make the impressions. Ultraviolet photos taken of the back image even show numerous fine scratches that would not be seen if the blood had not been washed from the body. This mechanism was easily demonstrated by briefly washing the wounds containing dried or clotted blood of victims of traffic accidents.

The victim would fall to his knees with each lash, writhing in agony, getting up each time until he could no longer lift himself up. There would be convulsive activity, tremors, vomiting, and marked thirst. Episodes of fainting would be associated with this type of flogging. The pain is so severe that many have pleaded for mercy and crying would be common. Periods of severe sweating would occur, intermittently. The severe pain associated with injuries of this degree would be a harbinger of traumatic shock soon to ensue and the fluid loss from excessive sweating coupled with the vomiting [and] the blood loss and sweating from the hematidrosis would cause an early stage of hypovolemia. The severe beating of the chest wall transmits to the lungs and promotes the gradual development of fluid around the lungs (pleural effusion), generally a few hours following the injuries.

THE CROWNING OF THORNS was not only a parody of Jesus' kingship but was another physical torture inflicted on Jesus. The tortuous flows on the forehead and the significant amount of blood on the head region had to have been the result of penetration of the skin by sharp thorns from a plant like those of *Ziziphus spina christi* (Syrian Christ thorn) or *Zizyphus paliuris christi* (Christ's thorn) both of the Buckthorn family (*Rhamnaceae*). Leading botanists of the plants of the holy land like Evanari, Post, Hegi, Tristram, Warburger, Moldenke, Schwerin and even the great Linnaeus were of the opinion that one or the other of the Ziziphus species were the most likely candidates. None of them even considered Gundelia tournefortii which has recently been implicated. Whether this plant is capable of penetrating the skin and inducing sufficient bleeding must be tested. From a forensic point of view, *Ziziphus spina christi* (Syrian Christ thorn) or *Zizyphus paliuris christi* (Christ's thorn) would cause puncture-type wounds with significant bleeding when struck with the reed ("and took the reed and struck him on the head" Mt.27:30) accounting for the blood flows and accumulations of blood in the head region of the Shroud.

It is of interest that the thorny acacia (*Acacia niltotica*) that grows profusely around the hills of Jerusalem has recently emerged as a contender. A crown of thorns made from this plant was unearthed in a sarcophagus dating to 1189 A.D. which also contained the remains of a mummified "knight of the temple" with a bashed skull and an inscription saying "this man saved the crown of thorns from the hands of the infidel." The physical effects of the crowning with thorns using a thorn plant like *Zizyphus paluris christi* as an example with its sharp, closely spaced thorns would most likely cause trigeminal neuralgia (tic douloureux) due to irritation of the ophthalmic branch of the trigeminal nerve (fifth nerve) and branches of the greater occipital nerves which supply sensory innervation to the front and back of the head region, respectively. This is characterized by severe, lancinating, paroxysmal, electric shock-like pains across the face lasting from seconds to minutes with intermittent refractory periods. Trigger zones are common in various areas of the face which trigger episodes of shooting pains across the head region if touched and is difficult to treat medically. Severe cases may not respond to medical treatment with drugs such as carbamazepine requiring nerve blocks or ablation surgery. The severe pain would be added to the depth of imminent traumatic shock now developing from the scourging. ...

THE CRUCIFIXION: Upon Jesus' arrival at Calvary, He exhibits a pale, mask-like appearance, is extremely weak, has severe thirst and his whole body is wracked with pain. He is in an early stage of traumatic and hypovolemic shock. After casting lots for His garments, they would have forced Him to the ground on His back, the patibulum placed just under His shoulders and upper back and members of the quaternio laying on top of Him to hold Him down and stretching out His arms on the patibulum while they drove iron spikes through His hands into the patibulum. This maneuver in holding Him down would cause almost unbearable pains in His chest because of the trauma from the scourging. It is well known in emergency medicine that trauma to the chest causes severe pain with the slightest pressure on the chest wall and with breathing. ...

The medical effects of the nailing of the hands whether it be through **the Z-area** or through the radial side of the wrist, would be essentially the same. The median nerve would be injured in either instance causing a painfully disabling affliction of the median nerve called *causalgia.* Causalgia can also occur in other peripheral nerves. The first full description of causalgia was described in 1864 by Mitchell, Morehouse and Keene[14] in reference to Civil War injuries. The pain in median nerve causalgia is an unbearable, exquisite pain described as a searing, burning unrelenting pain traversing the arms like lightning bolts. The person is unable to bear even the gentlest local contacts. It may be aggravated by movement, jarring, noise, a breeze or emotion. Increases in the ambient temperature or exposure to the sun would bring on more pain. Periodic episodes of marked sweating would also be manifested. The concomitant presence of fatigue greatly aggravates the degree of pain. Strong narcotic pain killers proved to be ineffective in many cases thereby requiring surgery to section the sympathetic nerves. Victims of causalgia frequently went into shock if the pain could not be controlled. This pain would have added significantly to the traumatic shock that was already in process.

The act of lifting the patibulum with Jesus' hands nailed to it in order to place it in a mortise at the top of the stipes that was anchored in the ground, would bring on renewed burning, and lancinating pains traversing the arms due to the pull of the hands against the nails. The hot temperature and exposure to the sun would increase the pain further. The pain was brutal, markedly increasing the degree of traumatic shock.

Next, the feet were nailed to the stipes by bending the knees in order to lay the soles flat to the stipes or one foot on top of the other and driving the spike through the feet. Branches of the medial plantar nerves would be injured affording pains of causalgia, similar to those of the hand described above. ...

DISCUSSION In order to arrive at the most probable cause of death, it is essential to examine the sequence of all the events from Gethsemane through Calvary; the severe mental anguish exhibited in the Garden of Gethsemane would cause some loss in blood volume both from sweating and hematidrosis and provoke marked weakness. The barbaric scourging that utilized a flagrum composed of leather tails containing metal weights or bone at the tip would cause penetration of the skin with trauma to the nerves, muscles and skin reducing the victim to an exhausted, wretched condition with shivering, severe sweating, frequent displays of seizures, and a craving for water. The results would cause a significant degree of trauma with impending shock (traumatic shock) and fluid loss and impending hypovolemic shock (fluid loss shock), the latter resulting from the various sweating episodes, and from the fluid accumulation around the lungs (pleural effusion) from the scourging. Animal experimentation by Daniels and Cates showed that blows to the chest in animals resulted in rupture of the air spaces in the lung (alveoli) and spasms of the air tubes (bronchi). Moreover the term "traumatic wet lung" refers to the accumulation of blood, fluid and mucus from severe trauma (injury) to the chest. This would be manifested several hours after the scourging. It may be of interest that the conclusion of traumatic shock from scourging, was also made by both

Tenney and Primrose. The irritation of the trigeminal and greater occipital nerves of the scalp by the cap of thorns especially after he was struck several times with reeds would also contribute to traumatic shock. The bumpy, uphill road to Golgotha in the hot sun, would incite trigger zones to initiate episodes of severe lancinating pain across the face due to trigeminal neuralgia and the carrying of the crosspiece on the shoulder for a time, with episodes of falling, also added to the oncoming traumatic shock and hypovolemia. The progression of the pleural effusion due to the scourging would lead to increasing hypovolemia. The large square iron nails driven through both hands into the patibulum would damage the sensory branches of the median nerve resulting in one of the most exquisite pains ever experienced by anyone and known medically as *causalgia*. The nails through the feet would also elicit severe pain due to causalgia from the injury to the plantar nerves. The causalgia would be aggravated by the sun, heat and fatigue. all of which would cause additional traumatic shock and hypovolemia. The hours on the cross, with pressure of the weight of the body on the nails of the feet and the pull on the hands would cause episodes of excruciating agony every time the *cruciarius* moved. These episodes of unrelenting pains added to the pains of the chest wall from the scourging would greatly increase the state of traumatic shock and the excessive sweating induced by the ongoing trauma and by the hot sun, would cause a increase in the degree of hypovolemic shock.

The pathophysiological events that occur as a result of these events leading to death are those of traumatic and hypovolemic shock. Shock, regardless of its cause is defined '... as a constellation of syndromes all characterized by low perfusion and circulatory insufficiency, leading to an imbalance between the metabolic needs of vital organs and the available blood flow.' It is '... a state of inadequate perfusion of all cells and tissues, which at first leads to reversible hypoxic injury, but if sufficiently protracted or grave, to irreversible cell and organ injury and sometimes to the death of the patient.' This presents a very complex array of initiating factors, compensatory reactions and several other interrelationships.[19]

Some Shroud skeptics (none of them medical professionals) have said that the scourge wounds are too "perfect" to be authentic. Apart from ignoring various medical studies on the matter, they overlook a few salient facts.

The Romans who administered the scourging would have to ensure that the intended victim stayed alive so the scourging could not be lethal. As trained, disciplined professionals, there would be nothing arbitrary or haphazard about their mode of inflicting punishment. This meant systematically "spreading it around" so that no one area of the body would be immobilized. The Gospels show Jesus carrying the cross which meant that he still had some degree of strength to do so after the scourging. But the fact that they had to find someone to assist him meant that he had been severely weakened.

Dr. Antero de Frias Moreira, a medical doctor, points out that "It's simply not true that all scourge marks have perfect dumbell shapes matching the Roman flagrum. It's impossible for the punisher to apply the same strength and direction to every lash, so some lashes would produce only contusions and bruisings of soft tissues while most of them would probably injure the subcutaneous fat and even muscles below and skin tearing could occur. When subcutaneous fat tissue and muscles were injured by the *flagrum plumbatae* bleeding would occur and soon an extended area of blood smear[ing] and clotting should follow. [The washing of the body after death becomes significant here.] By removing that large dried blood area by washing reopened scourging lesions that reached the subcutaneous fat and muscular layer [these] would bleed again even if their correspondent skin area was torn away and it is not unreasonable to assert that the shape of this re-bleeding would match the offensive object, because what bleeds is not the superficial skin layer but the subcutaneous fat and muscle; in other words the shape of blood marks from scourging is related to bleeding from muscle and fat NOT THE SUPERFICIAL SKIN. This is exactly what we observe on the Shroud ...

besides [the fact that] U.V. fluorescence photographs show a serum halo around the scourge marks."[20]

The reference to the serum halo is especially relevant. If the scourge marks were painted, they would have none of the hallmarks of blood. But chemists who have studied the Shroud have reached a different conclusion.

Robert Dinegar, a chemist from the Los Alamos National Scientific Laboratories, observed that "At the border of the scourge marks and around the periphery of the heavier blood flows, there is a white fluorescence, invisible in white light. When tested chemically this was identified as serum albumin. Blood degradation products in the form of bile pigments were also found."[21]

Another chemist, Kelly Kearse, wrote: "Numerous blood markings are present on the Shroud, which correspond to the wounds of a man that has been tortured and crucified. Bloodstains are evident about the head area, the wrists and the feet, the side, and the back. Each individual blood wound shows a distinct serum clot retraction ring; such blood halos are only visible under ultraviolet light, a detail that a forger is unlikely to have been familiar with. Over ten different chemical tests have established that these markings are indeed bloodstains, and contain specific blood components, including hemoglobin, bilirubin, and albumin. These data have been extensively reviewed elsewhere."[22]

The doctors' diagnoses continue. Recently, the journal *Injury*, the official publication of national trauma care associations around the world, published a paper by four Italian scientists on the suffering and death of Jesus as described in the New Testament and imaged on the Shroud The scholarly paper "How was the Turin Shroud Man crucified?" was "based on both image processing of high resolution photos of the TS and on experimental tests on arms and legs of human cadavers." The authors conclude that the Man of the Shroud "suffered the following tortures during crucifixion: a very serious and widespread causalgia due to total paralysis of the upper right limb (paradoxical causalgia); a nailing of the left wrist with damage to the ulnar nerve; a similar nailing of the right wrist; and a nailing to both feet using only one nail that injured the plantaris medialis nerves. The respiratory limitation was probably not sufficient to cause death by asphyxiation. Also considering the hypovolemia produced by scourging and the many other tortures detectable on the TS, the principal cause of death can be attributed to a myocardial infarction."[23]

The Italian newspaper *La Stampa* gave a popular-level overview of the study:

> The first discovery the four experts made, is that the Man of the Shroud underwent a dislocation of the shoulder and paralysis of the right arm. The person whose figure is imprinted on the Shroud is believed to have collapsed under the weight of the cross, or the "patibulum" as it is referred to in the study, the horizontal part of the cross. The Man of the Shroud the academics explain, fell "forwards" and suffered a "violent" knock" "while falling to the ground." "Neck and shoulder muscle paralysis" were "caused by a heavy object hitting the back between the neck and shoulder and causing displacement of the head from the side opposite to the shoulder depression. In this case, the nerves of the upper brachial plexus (particularly branches C5 and C6) are violently stretched resulting in an Erb-Duchenne paralysis (as occurs in dystocia) because of loss of motor innervation to the deltoid, supraspinatus, infraspinatus, biceps, supinator, brachioradialis and rhomboid muscles." At this point it would have been impossible for the cross bearer to go on holding it and this brings to mind the passage in the Gospel which describes how the soldiers forced Simon of Cyrene to pick up Jesus' cross. Not an act of compassion therefore, but of necessity. This explains why "the right shoulder is lower than the left by 10±5 degrees" and "The right eye is retracted in the orbit" because of the paralysis of the entire arm, the academics say.
>
> The second discovery described in the *Injury* article is to do with the double nailing of the Man's hands: Until now, experts could not explain the absence of thumbprints. The four academics can now reveal that "the lack of thumbprints of both

hands on the TS is related not only to a lesion of the median nerve that causes only a slight flexion of the thumb, but also, particularly, to the fact that the nail driven into the wrist has pulled or injured the flexor pollicis longus tendon causing its dragging in the hole and the complete retraction of the thumb."

Why the double nailing? One plausible reason could be that the Man's executioners were unable to nail his hands into the holes that had already been specially punched into the cross to prevent the nails from bending when they were hammered into the hard wood. Once the first wrist was nailed to the cross they failed to nail the second one using the pre-prepared hole and so the executioners had to unnail both wrists. They then apparently drove the nails in lower down between the two rows of carpal bones, on the ulnar side of the hand.

The third discovery is to do with the right foot of the Man of the Shroud: it was nailed to the cross twice. An analysis of the imprint of the sole of the right foot shows two nails were driven into it: one between the second and third metatarsal and another at heel level which other academics had not spotted clearly.

According to the four experts, the Man of the Shroud definitely suffered a very serious and widespread pain accompanied by an intense sensation of heat, and usually shock when there is even the slightest limb movement. This was caused by a total paralysis of the right arm, the nailing of the left arm because of damage to the median nerve and the nailing of the feet because of damage to the tibial nerves. This method of nailing led to breathing impairment: with the arms raised at an approximately 15 degree angle causing the ribcage to expand, the lungs had difficulty expiring, reducing air flow. In addition to this, each deep breath the Man took to speak or to catch his breath will have put a strain on the lower limbs causing him intense pain.

According to the authors of the *Injury* article, the serum stains, which are separate to the stains of blood that came from the chest and were probably caused by the stabbing with a spear after he had died, were formed as a result of bleeding in the lungs. This bleeding will have started before the crucifixion, after the violent fall which caused the cross to fall onto the Man's shoulders.

Finally, the authors of the article put forward their theory on the Man of the Shroud's immediate cause of death. Restricted breathing and the presence of the haemothorax which put pressure on the right lung were not enough to bring about death by asphyxia. Asphyxia involves an inability to breathe which results in loss of conscience and coma. The four experts say the fall and/or the flagellation have caused not only a pulmonary contusion but also a cardiac contusion. This, together with the serious clinical and mental condition the Man was in, may have led to a heart attack and a broken heart.

In their conclusion, Bevilacqua, Fanti, D'Arienzo and De Caro write that "from correspondences here and elsewhere detected between TS Man and the description of Jesus's Passion in the Gospels and Christian Tradition, the authors provide further evidence in favour of the hypothesis that TS Man is Jesus of Nazareth."[24]

La Stampa (January 2019) also reported another recent discovery: Filippo Marchisio, head of Radiology at an Italian hospital, and Pier Luigi Baima Bollone, professor of Forensic Medicine at the University of Turin, used CT scans to confirm that the Man of Shroud was crucified. They also identified the entry point of the lance.

As detailed by Diane Montagna,

The investigation started from the observation that the Man of the Shroud's right arm appears to be six centimeters longer than the left one. The two scholars attribute this apparent anomaly to a fracture in the elbow or a dislocation in the shoulder, which is compatible with a crucifixion. They also took into account that the arms would have had to be forcibly bent to overcome the stiffness of the body at the time of burial. ...

'The CT scan allows a perfect reproduction of the body, allowing us to reconstruct the missing parts without the subjectivity inherent in artistic creation,' Marchisio said. 'The CT underscores the inconsistency of the position of the shoulders and hands, a further element that supports the hypothesis that the man of the Shroud was really crucified.' ...

The researchers also identified the exact point at which a lance pierced the victim's side.

They were thus able to identify the organs that were injured, "releasing an accumulation of blood in the pleural cavity," i.e. the thin fluid-filled space between the two pulmonary pleurae of each lung.

Marchisio explained that "the blood mainly ran down on the right side, along a the channel formed by the arm contiguous to the body up to the elbow, and then it collected to form the belt of blood in the lumbar region."

"The anatomical relations revealed by the reconstruction of the missing parts confirm it: it is the demonstration of the extraordinary nature and the coherence of the Shroud," the researcher said. "The more one studies it, the more surprises it holds."[25]

As we said, the clinical study of the Man of the Shroud is part of the practice of medicine. The work of the professionals who have employed their expertise in this regard is an enduring contribution to our understanding of the Shroud.

2.4 Bloodstains

Inextricably tied to the doctors' diagnoses is the blood on the burial cloth. Like everything else connected to the Shroud, the actuality of the bloodstains has also been fiercely questioned. Skeptics are unmoved by the fact that blood chemistry authorities have quantified and demonstrated the presence of blood on the Shroud. So how can the matter be settled? Yet again, by listening to the researchers who have specialized in this domain rather than lending our ears to every voice crying out in the wilderNet. We are far more likely to solve a problem in nuclear physics by consulting a nuclear physicist than by roaming the dark Web. Would you rather rely on your Internet findings to create overload protection for your electrical system than hand over the problem to a certified electrician? Yes, even electricians make mistakes but these are unlikely to be of the elementary variety.

On the matter of blood, there has been only one hands-on professional study of the Shroud: the STURP investigation carried out by multi-disciplinary scientists from diverse belief backgrounds. In their final 1981 Report, they declared that "We can conclude for now that the Shroud image is that of a real human form of a scourged, crucified man. It is not the product of an artist. **The blood stains are composed of hemoglobin and also give a positive test for serum albumin**." This is the conclusion reached by the only specialists who physically handled the Shroud itself in analyzing its bloodstains and body image. Everything else is speculation. The radiocarbon dating study is irrelevant here because it had nothing to do with the provenance of the bloodstains. And, in any case, it was performed on a non-image section of the cloth.

STURP team members have provided a helpful overview of the work on the bloodstains:

The 'blood' areas on the Shroud have attracted considerable attention since the first color photographs of the cloth became available. It appeared that blood had flowed from the man's feet, wrists, and side.... The reddish, brown stains appear to be quite anatomically correct, as one would expect if a man had bled after being stabbed in the side and nailed through his wrists and feet. The edges of these stains are also precisely defined. If the Shroud actually covered a real corpse, one wonders how the cloth was removed without smearing and dislodging the edges of the clotted blood. When they arrived in Turin in 1978, the scientists did not know whether the 'bloodstains' were really blood. ... The 1978 team hoped to settle the blood question once and for all by examining the bloodstained areas with a full battery of optical tests throughout the electromagnetic spectrum. ... The most important and conclusive work was done by John Heller and Alan Adler in their laboratory at the New England Institute. [Heller, J.H. & Adler, A.D., "Blood on the Shroud of Turin," Applied Optics, Vol. 19, 1980, pp.2742-2744] Heller and Adler examined several 'sticky tape' samples which contained pieces of 'bloodstained' fibrils. They looked at the spectrum of the visible light transmitted from these samples under a microscope, a test known as microspectrophotometry. The results suggested that hemoglobin was a component of the color. To further test this possibility, Heller and Adler removed the iron from the samples and tried to isolate porphyrin, a component of blood which fluoresces red under an ultraviolet light. Indeed, the substance which the chemists isolated from the samples fluoresced red under ultraviolet light. This confirmed that the substance was porphyrin, and thus strongly indicated that the bloodstained areas really were blood. A further indication that blood was present on the Shroud came from the ultraviolet fluorescence photographs taken by Vernon Miller and Samuel Pellicori. Blood itself does not fluoresce. However, when Miller and Pellicori studied their ultraviolet fluorescence photographs of the blood areas, they discovered a light fluorescent margin around the edges of several of the bloodstained areas. These areas were the side wound, the nail wound in the wrist, and the blood flow at the right foot on the dorsal image. The probable explanation for this unexpected discovery is that the fluorescent margins were blood serum, the colorless fluid part of the blood. Miller and Pellicori showed in the laboratory that blood serum on linen does fluoresce moderately. Thus, it is likely that the fluorescent margins are blood serum which had become separated from whole blood before or after the man's death. Several other tests confirmed the presence of blood on the Shroud. Protein, a component of blood, was detected in the blood areas, although no protein was found elsewhere on the cloth. X-ray fluorescence examination found that iron, a component of blood, was present in the blood area. The team's summary of research concluded that the bloodstained areas were very probably stained by real blood.[1]

The STURP blood chemistry specialists, John Heller and Alan Adler, had the specialized expertise for the task. As earlier highlighted, Heller was a professor of Internal Medicine and Medical Physics at Yale University. He was the founder of the New England Institute, the first and (initially) only research institution dedicated to the study of host defense systems. Heller also founded the Reticuloendothelial Society and started the RES Bulletin that is today the prestigious *Journal of Leukocyte Biology*. In addition to his research renown, he was called on to test for the presence of blood in murder cases. Adler was the author of several hundred scientific papers in chemistry and biochemistry with a focus on porphyrins, biological pigments that are essential for the function of hemoglobin. Adler, who was Jewish as we noted, had never even heard of the Shroud of Turin when he was first approached to participate in the STURP project.

Because of his faculty position, Heller was able to use the microspectrophotometer at Yale University to analyze the blood flecks in the Shroud samples. In all Heller and Adler performed over one thousand chemical tests and published a paper on their results, titled

"A Chemical Investigation of the Shroud," in the peer-reviewed *Journal of the Canadian Society of Forensic Sciences*.

The twelve main categories of tests performed by Heller and Adler each confirmed the presence of whole blood on the Shroud as reported by them (see below) in their Forensic Sciences paper[2]:

1) High Fe (Iron) in blood areas by X-ray fluorescence
2) Indicative reflection spectra
3) Indicative microspectrophotometric transmission spectra
4) Chemical generation of characteristic porphyrin fluorescence
5) Positive hemochromagen tests
6) Positive cyanmethemoglobin tests
7) Positive detection of bile pigments
8) Positive demonstration of protein
9) Positive indication of albumin specifically
10) Protease tests, leaving no residues
11) Microscopic appearance as compared with appropriate controls
12) Forensic judgement of the appearance of the various wound and blood marks

This is how they summarized the takeaway from these tests: "That means that the red stuff on the Shroud is emphatically, and without any reservation, nothing else but B-L-O-O-D!"[3]

It has been pointed out that "it only takes three out of the twelve tests to prove the presence of whole blood in a court of law. On the Shroud there are four times that number." Further, "The high presence of iron on the Shroud is significant, as iron is a major component of blood. Also noteworthy is the presence of bile residue known as bilirubin. This occurs when blood begins to break down, particularly after someone has suffered a severe trauma."[4]

No one before or since has studied the bloodstains in such detail. Heller was emphatic that all their research should be published in leading peer-reviewed publications and this is indeed what happened. While anyone with an Internet connection can second-guess the implications of this phalanx of data, nobody has been able to generate a serious refutation of the work in a peer-reviewed publication.

And this brings us to an amusing if diversionary episode in the study of the blood images: the antics of microscopist Walter McCrone. Although McCrone told anyone who would listen that he was "drummed out" for being a dissenting voice, it is a matter of record that he could not get any of his theses on the Shroud published in a peer-reviewed publication – let alone have his tests replicated. Neither was he willing to address criticisms of his work from fellow-scientists. And he declined several opportunities to defend his positition in face to face dialogues, e.g., turning down an invitation from the Canadian Society of Forensic Sciences to debate Shroud scientists in Hamilton, Ontario.

Of course, there was an additional problem: McCrone kept changing his mind on what was wrong with the Shroud! Worse yet, even if he were right in affirming that there was vermilion or iron oxide on the Shroud this did not mean that there was no blood on the Shroud. It simply confirmed matters that were already known, e.g., various paintings had been laid on the Shroud to be "sanctified" and some exchange of particles from these paintings was to be expected.

Thomas de Wesselow points out that "there is no sign of any pigment or binding medium" on the Shroud. "McCrone based his interpretation of the Shroud as a painting on his observation of certain 'pigment particles' on the fibers of the cloth, but these are best interpreted in other ways. Particles of iron oxide (the main constituent of "red ochre") can indeed be observed on and in the fibers of the cloth, but they are not the cause of the image. Unlike pigment particles, they are tiny and very pure and must have formed during the process of manufacture, when the cloth was retted (i.e., soaked) in iron-containing water. The odd particle of vermilion, on the other hand, which McCrone linked to the blood image, should be understood as contamination, probably derived from the painted copies laid on the Shroud in centuries past. ... Tellingly, he failed even to convince his fellow skeptics."[5]

One such fellow-skeptic, Harry Gove, observed that McCrone had already developed a reputation for jumping to sensational conclusions without the requisite evidence:

I sometimes think that McCrone dreamed of becoming history's greatest iconoclast. Having, in his view, demolished the authenticity of the Vinland Map he saw the chance to do the same to the Turin Shroud! When a series of tests were carried out on the shroud in the fall of 1978, McCrone determined that there were traces of iron oxide powder on the shroud image.[6]

More recently, doubts have been raised concerning McCrone's conclusions and there is increasing evidence that the Vinland Map predates Columbus' discovery of America.[7]

The problem McCrone has is that his scientific techniques are unsophisticated.[8]

In an article in the *Times Literary Supplement*, Archaeologist John Ray of Cambridge University has given a sympathetic account of McCrone's lost labors:

The bravest attempt to declare the Shroud a painting was made by Walter McCrone, a respected art investigator. McCrone detected red particles on some of the fibres containing the image, and argued that the image was a simple product of red ochre or vermilion and an adhesive. ... This explanation is superficially attractive, but it fails to account for some important details. The suggested adhesive has not been identified on the Shroud, and the particles, which certainly exist, appear to cling to most of the surface of the cloth, rather than to be closely associated with the image. High-resolution magnification suggests that, however the image was produced, it was not done this way... Another problem is that McCrone's artist [hired to replicate the Shroud], who was highly competent, nevertheless produced a work which looks like a painting. He did this working from excellent, full-scale photographs of the Shroud. What was the original mediaeval artist working from? All he (and we are surely justified in thinking this was the artist's gender?) could have seen as he worked is the faint positive image visible to the eye. Within a distance less than ten feet or so from the cloth, the image disappears, so he would have needed to stand well back from his work to see it. ... How did our shifty depictor in Lirey produce a negative image, which he could not control, and whose anatomical accuracy appears to be faultless, without once giving away the fact that he was painting?[9]

McCrone was unwilling to accept results from newer testing technologies: "Trained on, familiar with, and devoted to the polarized light microscope, McCrone was reluctant to do wet-chemistry testing and loathe to accept the peer-reviewed results of the 1978 physics-based testing." Heller and Adler repeatedly responded to McCrone's claims but McCrone simply failed "to respond in print to contrary peer-reviewed data and conclusions."[10]

In a 1993 interview Adler recounted, "We pointed out that, yes, we saw what [McCrone] claimed he saw. We saw iron oxide, we saw one piece of vermilion, we saw protein. We also saw red particles that weren't iron oxide. ... The red we saw was blood, and it was in the blood tape samples from Turin – the only places we saw iron oxide was in the water stain areas and the blood scorch areas. And our explanation is, when you burn blood you get iron oxide – it contains iron."[11]

Some critics accept the reality of the blood on the Shroud but then suggest it was added in later years. Adler points out that this is simply not possible while also reviewing the evidence for the blood:

> Although they sometimes differ on certain matters, all of the medical forensic examinations of the blood images are in agreement that they were exudates from clotted wounds transferred to the cloth by its being in contact with a wounded human male body consistent with the historic descriptions given for the Crucifixion of Christ. This conclusion is also consistent with the computer imaging evidence. A simple masking transfer experiment has shown that the body images are out of stereoregister with the blood images and therefore

have gotten onto the cloth by a non-contact information projective process. This is in agreement with the original observations of Vignon and the more recent computer imaging studies. Enzymatic removal of the blood from a blood coated fiber reveals that the blood got on the cloth first and therefore protected the blood covered areas of the cloth from the image forming process. All the microscopic, chemical, spectroscopic, and immunological evidence is consistent with these images, not only being exudates from clotted wounds, but those of a man who suffered severe trauma prior to death, explaining the red color of the blood at the microscopic level. Proposed mineral compositions simulating blood are not consistent with these various measured chemical and physical parameters. That these are clotted wound exudates is clearly seen in the ultraviolet photographs where every single blood wound shows a distinct serum clot retraction ring agreeing with the earlier observations of the pioneers on the major blood wounds as seen directly on the cloth. It is clear that we can explain the presence of the blood images on the cloth consistent with their alleged origin.

Note that any attempt to explain the formation of the body images must take these properties of the blood images into account. One cannot simply say that the blood images were painted on afterwards. One would need a constant supply of fresh clot exudates from a traumatically wounded human to paint in all the forensically correct images in the proper nonstereo register and then finally paint a serum contraction ring about every wound. Logic suggests that this is not something a forger or artisan before the present century would not only know how to do, but even know that it was required.[12]

The work of Heller and Adler was complemented by the experiments of Pierluigi Baima Bollone, an Italian professor of Forensic Medicine. Where Heller and Adler used samples secured by adhesive tapes, Bollone worked with threads directly removed from the cloth. Using high-end equipment and relevant methodologies, Bollone was able to detect the presence of diverse components of blood. In his paper, "The Forensic Characteristics of the Blood Marks," Bollone said, "Under the fluorescence microscope and using the Dotzauer and Keding method on the same samples I demonstrated the presence of heme/porphyrins. On the same material I obtained Teichmann crystals or hematine chlorohydrate with the usual procedures. ... After optical and scanning electron microscope investigations, I managed by means of the energy extinction microspectrometer to ascertain the presence of Mg, Al, Si, S, Cl, K, Ca and Fe."[13]

As in other areas of Shroud studies, blood image researchers continue to publish new works. One such researcher, Kelly Kearse, wrote that

> Careful conclusions about the blood on the Shroud (and related artifacts) must rely on multiple approaches that support and confirm each other. It is somewhat precarious to base deductions on a single type of evidence. The use of techniques and methodologies that crosscheck and verify specific findings are instrumental in any type of scientific investigation. Ideally, additional serological analyses with greater sensitivity and DNA studies of specific coding segments (for example, ABO, S, HLA transplantation antigens) could be included among such data on the Shroud in the future. In summary, the preponderance of current scientific evidence indicates that: (i) there is blood on the Shroud of Turin; (ii) the blood is of primate, i.e. human origin.[14]

Most recently, the first ever nanosize study of a fiber from the Shroud, performed by two prestigious European scientific centers and published in the peer-reviewed journal PLOS, settled the painting vs. blood question. The study concluded that the blood image on the Shroud is not dye from a painting but blood serum from a tortured person. The journal paper titled *Atomic resolution studies detect new biologic evidences on the Turin Shroud* points out that the team performed reproducible atomic resolution Transmission Electron Microscopy and Wide Angle X-ray Scanning Microscopy experiments studying for the first time the nanoscale properties of a pristine fiber taken from the Turin Shroud. ... On the basis of the experimental evidences of our atomic resolution TEM studies, the man wrapped in the TS suffered a strong polytrauma. ...The fiber was soaked with a blood serum typical of a human organism that suffered a strong trauma. ... The obtained results are not compatible with a painting but evidenced the presence of nanoparticles of pathologic blood serum related to the presence of creatinine bound with ferrihydrate, which are typical of an organism that suffered a strong polytrauma, like torture. Indeed, unexpectedly, at the nanoscale it is encoded a scenario of great suffering recorded on the nanoparticles attached to the linen fibers.[15]

2.5 "The stones will cry out" – Pollen, Dust, Cloth

Unexpectedly, various fellow-travelers of the Shroud have burst out of the blue and point to its first century Middle Eastern origin. These include the pollen and dust particles on the Shroud, and the design and material of the Shroud itself. More controversially, there is the claim that the Shroud displays Jesus' "death certificate!"

Pollen

The pioneer in exploring the pollen on the Shroud was the Swiss botanist Max Frei. With the support of Church authorities, Frei secured dust from the Shroud and announced his findings in 1976:

> The first conclusion that the performed studies allow to suggest refers to the presence on the Shroud of pollen grains that come from desert plants that grow in Palestine. The most frequent pollen on the Shroud is identical to the most frequent pollen in sediments of the lake of Gennesaret sedimentary layers of two thousand years ago. Another sample comes from Asia Minor and more specifically from the area surrounding Constantinople, while a large number of granules are of French and Italian origin. It is therefore logical [to deduce] that the geographical and historical life of the Shroud corresponds to the migration that it suffered in time as a function of the evidence acquired.[1]

Commenting on this announcement, Emanuela Marinelli writes,

In the twelve dust samples, Frei found, in addition to the pollen of flowering plants, fiber fragments, mineral particles, fragments of plant tissues and fungal spores.

With regard to pollen, he reminded: 'Every species of plant produces a unique pollen that can be distinguished from the pollen of all other varieties, both under the light microscope and under the scanning electron microscope. (...) It is then possible to determine on the basis of a single grain of pollen from which plant it comes.' He also stated: '90% of the pollen production of a given plant is deposited within 100 meters. A propagation to a distance of tens of kilometers is still considered normal, while exceptionally strong winds in times of drought (sirocco) are responsible for rare extremely far transports of hundreds or thousands of kilometers. (...) In the case of the Shroud the represented plants bloom in different seasons and live in well-defined, and different from each other, ecological conditions. Their pollen is not especially suited to very far transports. Therefore the heterogeneity and the amount of pollen cannot be explained on the basis of random contamination.'"[2]

Later botanists differed significantly from Frei's identification of the most common species of plant pollen on the Shroud. But his pioneering work opened the door to a new way of studying the Shroud. The most recent research reports on the botany of the Shroud are distinctive in that they appear in preeminent peer-reviewed scientific journals – and support a Middle Eastern origin for the Shroud.

Most significant of these is an October 2015 supplement published by *Nature* (the journal associated, of course, with the Shroud carbon 14 report) and titled "Uncovering the sources of DNA found on the Turin Shroud." Remarkably, the paper, authored by four scientists, notes that recent research studies "have highlighted some concerns about this determination [of a medieval origin of the Shroud] and "a

Medieval age does not appear to be compatible with the production technology of the linen nor with the chemistry of fibers obtained directly from the main part of the cloth in 1978." The paper states:

> In 1978 and 1988, dust particles were vacuumed from the interspace between the Shroud and the Holland Cloth sewn to it as reinforcement. The composition of the particles was later studied in great detail by optical microscopy, and specimens from different filters were retained and characterized for their contents. In past decades, pollen grains were classified to the genus and species levels using microscopy, and the geographic areas where the corresponding plants originated and now inhabit proved to be compatible with the reported historic path followed by TS during the postulated 2000-year journey from the Near East, thus supporting the authenticity of the relic. ...
>
> In this study, we performed DNA analyses to define the biological sources of the dust particles (pollen grains, cell debris and other minuscule organic specimens, such as plant-derived fibers and blood-like clots) vacuum-collected in 1978 and 1988 in distinct TS filters, corresponding to the face, hands, glutei and feet of the body image, and the lateral edge, which was used for radiocarbon dating. ...
>
> With regard to the land plant species identified, some are native to Mediterranean countries and widespread throughout Europe, North Africa and the Middle East and are thus compatible with both a rather recent Medieval origin in Europe and a more ancient Near Eastern origin."[3]

Another paper that appeared in the journal *Archaeometry* titled "Pollen on the Shroud of Turin: The Probable Trace Left by Anointing and Embalming" (2016) and authored by Marzia Boi links the Shroud's "pollen traces to a mixture of balms and ointments employed for preparing the body for funeral and burial."[4]

At a conference, Boi reported that: "The pollen traces on the Holy Shroud which have so far been linked to the geographic origin of the relic reveal what oils and ointments were put both on the body and on the sheet. These discoveries have an ethno-cultural meaning linked to ancient funeral practices. **These non-perishable particles capture the image of a 2000-year-old funeral rite and thanks to them it was possible to discover what plants were used in the preparation of the body that was kept in the sheet.** The oils allowed the pollens, as fortuitous ingredients, to be absorbed and hidden in the shroud's fabric like invisible evidence of an extraordinary historical event."

The newspaper *La Stampa* notes that "According to Jewish custom the dead bodies and the winding sheets were treated with oils and perfumed ointments following a meticulous ritual."

Boi's work led her to conclude that the main pollen residue on the Shroud is not Gundelia Tourneforti. "Her examination with the electron microscope yielded a different result: **the main pollen residue comes neither from Ridolfia, nor Gundelia, but from Helichrysum** (29.1%). Cistaceae pollen (8.2%), Apiaceae pollen (4.2%) and Pistacia pollen (0.6%) are also present on the shroud in smaller quantities. "All the plants mentioned here are entomophilous, that is, their pollen is carried by insects rather than air. This shows that there must have been direct contact with either the plants or the materials used for the funeral. The list of pollens reveals traces of the most common plants used in ancient funerals. The pollens identified clarify that the holy shroud was rubbed with oils and ointments, just as the body contained within it did." There used to be a balm made from Pistacia leaves, fruits and bark that was also used as an ointment. However, a high quality oil was once produced from the Helichrysum and this oil was used to protect both body and shroud."

"The use of this oil in ancient funeral rites is documented in various countries, from Arabia to Greece."

Boi concluded: "Identifying the main pollen traces found on the Shroud captures a snapshot of a funeral rite that followed the customs of Asia Minor, 2000 years ago. They are the components of the most precious oils and ointments of the time and have extraordinarily remained sealed in the fabric… The correct identification of the Helichrysum's pollen, wrongly believed to be that of the Gundelia flower, confirms and guarantees that the body wrapped up in the sheet was an important figure."[5]

Another peer-reviewed paper that appeared in the journal *Archaeological Discovery* in 2015 again indicates a Palestinian origin for the Shroud:

> We studied by SEM-EDX analysis the pollens on the Face of the Turin Shroud. A total of ten pollen grains were found; they were photographed, characterised and analysed. Three of them (pollens p6, p7 and p10) belong to Ceratonia siliqua, the carob tree; one of them (pollen p1) belongs to Balanites aegyptiaca (the palm tree of the desert), and another one (pollen p9) belongs to Cercis siliquastrum (the Judean tree). These three plants have their geographical distributions in the Near-East; that is indicative of a Palestinian origin of the Turin Shroud. Two pollen grains (p3 and p4) belong to Myosotis ramosissima. Probably myositis flowers were deposited later on the Turin Shroud, as reverence for this venerable and symbolic object.[6]

The work of two Israeli botanists Avinoam Danin and Uri Baruch has also been influential over the last two decades:

> Botanical experts on the research team found the imprints of plants and grains of pollen that can serve as seasonal calendar and geographic indicators.

> Four plants on the shroud are significant because, as researchers Danin and Baruch report, "the assemblage...occurs in only one rather small spot on earth, this being the Judean mountains and the Judean Desert of Israel, in the vicinity of Jerusalem."

> Those experts succeeded in identifying thirty-six species of plants on the shroud. They discovered that almost all of the flower images remaining on the cloth and the highest concentration of pollens were where the head of the corpus

would have been lying; plant parts and pollens were also located throughout the rest of the shroud.

The botanists found several factors of particular interest to those studying, even doubting, the authenticity of the shroud. These are some of their findings:

All the plants are ones that grow in Israel. Of these, twenty are known to grow in Jerusalem itself and eight others grow in the vicinity in the Judean desert or the Dead Sea area.

Although some of these plants are found in Europe, fourteen plants grow only in the Middle East.

Twenty-seven of the plants bloom in the springtime at the same time as the Jewish Passover.

Zygophyllum dumosum has both pollen as well as an image on the shroud and grows only in Israel, Jordan and the Sinai region.

Gundelia tournefortii ... was the plant material found where the Crown-of-Thorns was imprinted around the head on the cloth.[7]

Dust

Assuming the Shroud was the burial cloth of Jesus, its first resting-place would have been a tomb in Israel. Limestone particles on the Shroud, when tested in high-end laboratories, turned out to have the same unique signature as dust in the caves of Jerusalem.

"Joseph Kohlbeck and Richard Levi found travertine aragonite limestone particles in sample dust collected from the Shroud's surface. Kohlbeck was Resident Scientist at the Hercules Aerospace Center in Utah. Levi-Setti was with the Enrico Fermi Institute at the University of Chicago,

Using a high-resolution microprobe, Kolbeck and Levi-Setti compared the spectra of the dust from the Shroud with samples of limestone collected from ancient tombs in Jerusalem. They found that chemical spectral data were identical except for some minor bits of cellulous fiber that could not be removed from the dust.

This is not absolute proof that the Shroud was in Jerusalem, for there might be other places in the world where travertine aragonite has the identical trace chemical signature. However it is statistically unlikely any will be found.

Some skeptics have suggested that the limestone dust was added by a forger creating the Shroud in the Middle Ages. That is highly implausible. Would he have used limestone from the environs of Jerusalem in anticipation of modern forensic analytical tools.[8]

Two other accounts offer more detail on this discovery.

C.B. Ruffin:

Scientists found other interesting features connected with the Shroud. Joseph Kohlbeck, an optical crystallographer ... found particles of aragonite with small amounts of strontium and iron on the Shroud's fibers on the image of the foot. With the help of archaeologist Eugenia Nitowski, he obtained samples of limestone from inside ancient tombs in and near Jerusalem and subjected them also to microscopic analysis. He found the same substance. The aragonite on the Shroud and in the tombs was an uncommon variety, deposited from springs, typically found in limestone caves in Palestine, but not in Europe. The samples from the Shroud and the tombs provided 'an usually close match,' suggesting to him and to Nitowski that the Shroud had once been in one of the 'rolling-stone tombs' that were common in Palestine around the time of Christ and for several centuries before. Kohlbeck observed that those who believe that the Shroud is a forgery need to explain how the very rare aragonite found its way to the surface of the Shroud.[9]

Ian Wilson:

Kohlbeck took both the Shroud samples and the Jerusalem tomb samples to Dr Ricardo Levi-Setti of the famous Enrico Fermi Institute at the University of Chicago. Here, Levi-Setti put both sets of samples through his high-resolution scanning ion microprobe, and as he and Kohlbeck studied the pattern of spectra produced by each ... it became quite obvious that they were indeed an unusually close match, the only disparity being a slight organic variation readily explicable as due to minute pieces of flax that could not be separated from the Shroud's calcium.[10]

Cloth

As unique as the pollen and dust on the Shroud is the design of the cloth. Mechthild Flury-Lemberg, "a world authority on ancient textiles" and curator of a Swiss textile museum, was charged with the Shroud conservation efforts. She was thus one of the first experts to access the opposite side of the Shroud. To her surprise she found that the style of stitching on this side has been seen before in only one location: in the ruins of the famous fortress of Masada where the Jewish population had been massacred by the Romans in 74 AD.

Flury-Lemberg said at a conference that "the weave and style of the material were from the Dead Sea area and could only have been woven in the period from 40 years before the birth of Christ up to 70 years afterward. The material, a rare 3-to-1 herringbone twill weave of hand-spun linen, is so unique that 'there is no way it could have been a forgery from the 13th or 14th centuries.'"[11]

A Swiss publication recounts the history of Flury-Lemberg's interaction with the Shroud and her views on the implications of her discovery:

Internationally renowned Swiss textile conservator Mechthild Flury-Lemberg ... first came into close contact with the shroud in 2002 when she was asked to help conserve it. In the course of that work she made a very detailed analysis of the fabric ..

The shroud suffered its greatest damage in 1532, when the chapel it was kept in caught fire, and drops of molten silver from the case which held it burnt large holes in the fabric. Nuns repaired the shroud by sewing it onto a backing cloth and placing patches over the holes.

These patches remained untouched for over 450 years until Flury-Lemberg and her assistant removed them. Underneath they found accumulated carbon dust, which had also got into the backing and into the fabric of the shroud itself as it had been rolled and unrolled over the centuries to be exhibited to the faithful.

She believes this dust could very well have skewed the data used in the carbon-14 tests.

She also points to the corner that was used for the tests, which was part of a water stain. "When the water runs over the textile it takes all the dirt with it. So at this corner you have the whole concentrated dirt. You can imagine that it was an extremely dirty sample, and also because at the corners of course people always held it," she explained.

"Therefore it made no sense to analyse this piece using carbon-14."

It is one thing to cast doubt on the reliability of the test, but the textile historian is interested in clues that offer alternative hypotheses. For Flury-Lemberg, a manuscript from Constantinople known to date from the 1190s is a clear indication that the shroud was certainly in existence at the end of the 12th century.

"The painter of this document must have seen the Shroud of Turin and must have known it as the shroud of Christ," she says.

The Pray codex, now in the Széchényi library in Budapest, shows the dead Christ on a strip of white linen. His arms are crossed, and only four fingers are visible.

"This is only on the shroud. The whole painting tradition puts the nail marks on the palms," she points out. But in fact the nails used in the crucifixion would have been put through the wrists to hold the weight, forcing the thumbs inwards.

But the textile historian sees other indications that the subject of the illustration in the Pray codex is the item now known as the Turin shroud. One is the herringbone weave, a stylised version of which is clearly visible. The other is the pattern of L-shaped marks probably caused by oil soaking through many centuries ago when the cloth was folded.

"So you have two marks of the Turin Shroud. Marks which are very typical of the shroud in Turin are painted here. That's the other reason why I never trusted the Oxford result."

A similar herringbone pattern has been attested in weaving from long before the time of Christ. The seam that attaches a long strip along one side of the shroud is in a style that has been found on textiles found in the fortress of Masada in Israel which dates to around the beginning of the Christian era.[12]

Flury-Lemberg expanded on her analysis in an interview.:

A renowned textile historian has become the latest specialist to say that the Turin shroud bearing the features of a crucified man may well be the cloth that enveloped the body of Christ.

Disputing inconclusive carbon-dating tests suggesting the shroud hailed from medieval times, Swiss specialist Mechthild Flury-Lemberg said it could be almost 2,000 years old.

Perhaps even more important is what Mrs. Flury-Lemberg saw when she examined the back of the shroud, the first researcher ever to do so. While it bore bloodstains, there were no mysterious marks comparable to those on the front of the cloth.

These marks show an amazingly detailed picture of a bearded man who had been beaten about the body, crowned with thorns and pierced with nails through the wrists and the feet.

On the side of the body's outline there appeared to be an image of a wound, which was perhaps the one caused by a Roman soldier's spear when he tried to find out if the crucified Jesus was alive or dead (John 19:34).

It was to this fist-size wound the resurrected Jesus guided the apostle Thomas' fingers, whereupon the doubting disciple explained, "My Lord, my God." (John 20:28).

Mrs. Flury-Lemberg, a Hamburg, Germany-born scholar now living in Bern, Switzerland, did preservation work on the shroud this summer. She said the outline of the body looked somewhat like burn marks, but only in the top 2 millimeters of the cloth. ...

Mrs. Flury-Lemberg investigated the cloth this summer as she separated it from the Dutch linen cloistered nuns in Chambery in Savoy had sewn it to after a fire in 1534.

She explained the linen's progressing oxidization had been endangering the shroud. As she separated the two textiles, she removed "spoonfuls of soot." She cleaned the shroud before it was sewn to a new cloth.

Pollen analysis and the shroud's measurements suggested it originated in the Middle East and not in medieval Europe. Mrs. Flury-Lemberg described its quality as "stunningly noble, with an almost invisible seam."

She related she discovered identical forms of weaving and high-quality sewing on textiles found at Masada, the ancient fortress in southeastern Israel. They hailed from the year 73 A.D.

According to the Bern scholar, other first-century cloths found in the Red Sea region showed weaving patterns similar to those of the Turin shroud. ...

Mrs. Flury-Lemberg questioned the relevance of findings by other researchers who discovered pollen and dust traceable to the Middle Ages on the cloth.

"Of course it had such particles on it," she said, "after all, the shroud was exhibited a great deal in those days."

Historian Karlheinz Dietz of Wuerzburg University in Germany shares Mrs. Flury-Lemberg's doubts of the 1988 carbon-dating results claiming that the cloth was made between 1260 and 1290.

In an interview with the German daily, Die Welt, he stated, "If you believe that the cloth hails from the Middle Ages then you must also believe that a man looking exactly like Jesus was whipped, crowned with thorns, crucified and then placed on linen imported from the Middle East and sprinkled with aloe and myrrh, and that on top of all he had invented monumental photography."

Mr. Dietz was referring to the discovery of the Christ-like image by Italian photographer Secundo Pia in 1889.

"On the shroud we see a genuine 'photography' that originated long before photography was invented," he said.[13]

Shroud scientist Emanuela Marinelli has also spoken to the implications of the design characteristics of the Shroud:

> As for the artifact, the threads of the cloth were spun by hand with the "Z" twist widespread in Syrio-Palestine in the first century A.D. The weave of the fabric, of the "herringbone" sort, comes from a rudimentary treadle loom. It has jumps and mistakes in the passage of the spindle. Herringbone fabric is of Syrian or Mesopotamian origin. In the finds of Jewish fabrics in Masada in Israel a special type of hem, like that on the Shroud, is known for the period between 40 B.C. and the fall of Masada in 74 A.D. On the Shroud there is also a longitudinal seam, identical to that found on fragments of tissue from the finds at Masada. So the technique of manufacture and type of tissue give a dating consistent with the time of Christ. One can add that the dimensions of the cloth (even if the size of the object may also have varied significantly because of repeated displays, with resulting unrolling, rolling, stretching and pressing) seem to match whole numbers in Syrian cubits, a unit of measurement used in ancient Israel. Other systems of measurement seem to match less well, in terms of whole units, the length and width of the cloth. It is also interesting that the parts of the fabric of the Shroud that could be examined did not reveal traces of animal fiber, in conformity with the Mosaic law that requires wool to be kept separate from flax (*Dt* 22, 11), the only other (minimal) traces of other fibers found in the cloth are cotton of the type *Gossypium herbaceum*, widespread in the Middle East at the time of Christ.[14]

Much has been made of the discovery in Israel of a woolen shroud from the first century made of a different weave, a two-way weave. The shroud was recovered from the Jerusalem tomb of a man suffering from leprosy. The researcher who made the discovery notes, "The shroud is made of wool. The Z-spin of the wool suggests production outside of Israel as Z-spun threads form only a small proportion of textiles in Israel and its neighbouring countries in the Roman period. ... The use of wool textile in primary use for burials and shrouds is less common than linen in the Land of Israel and was usually used for shrouds in secondary use. Linen shrouds have been discovered at burial sites in the Land of Israel."[15] For reasons noted below, the discovery of this shroud has no particular bearing on the Shroud of Turin.

In private correspondence, Dr. Flury-Lemberg pointed out to us that

> A small tessera of this mosaic [the 1/3 herringbone twill weave of the Shroud] is represented by the mountain leggings that have been recently discovered after the melting of a snowfield on the Vredetta of Ries in the Trentino Alto Adige. The 1994 [discovery] concerns a find going back to the Iron Age and it is dated between 795 and 499 B.C. The leggings are made of raw goat wool and manufactured with a 2/2 herringbone [weave] (warp: colour beige, weave: colour brown and dark brown) They are without soles and are 54,6 cm long and 15,7 cm wide. They are made in a piece of fabric 31 cm width, with selvage to the two ends and these are bound by a seam in the inside. An interesting bond between the linen of Turin and the leggings is represented by the mutual pattern of the herringbone. This is a proof that we can find this technique of weaving since ancient times, even though in [the] case of the leggings it is a very rough woolen cloth. The woolen cloth was manufactured on a more primitive loom than the one of the Shroud of Turin, but the technique was already known in [the] pre-Christian era, and it is not limited to a specific region.

As the Gospel accounts have it, Jesus was wrapped in a linen cloth provided by Joseph of Arimathea, a wealthy man. It can plausibly be assumed that whatever material Joseph had would be of a high quality.

With regard to the Shroud itself, Dr. Flury-Lemberg makes a remarkable observation:

In spite of the high quality of the Turin cloth no doubt it will not have been produced for the use as a "grave linen." It may have been meant [for] "normal" usage, for instance as a "table cloth." Grave linen, that is linen to be buried with the body in the earth, had a [lesser] quality. The (for Israel) unusual structure ("heringbone twill cloth") as well as the high quality of the fabric looks as [if it were] meant for special use. It could be a precious import from Egypt or Syria [or] professionally manufactured in Israel (for instance Masada). The cloth probably was used to cover the body because it was valuable and was at hand at the right time. The body remained in the cloth obviously a very short time. All details point to a hasty and provisional use.

Death-certificate?

One intriguing but controversial development in Shroud studies is the claim that it displays a "death certificate" for Jesus the Nazarene. As far back as 1979, the Italian researcher Piero Ugolotti announced that digital image processing showed Greek and Latin letters in the facial area of the Man of the Shroud. These letters, it was claimed, were not visible to the naked eye because of their low contrast but showed up when photographic images of the Shroud were digitally processed. Ugolotti worked with Aldo Marastoni, a specialist in ancient languages, who claimed that (among other things) the inscriptions said "Nazarene." French researchers from the Institut d'Optique Theorique et Appliquee d'Orsay processed the images and presented it at a conference in Nice. It was further studied by another researcher from France, Thierry Castex. In 2009, Barbara Frale, a Vatican archive researcher whose specialties included Archaeology and Palaeography, published a book *The Shroud of Jesus the Nazarene*, expanding further on this claim.

In an interview with "La Repubblica," Frale said that "Jewish burial practices at the time of the Roman occupation of Jerusalem mandated that a body buried after execution of a death sentence had to be in a common grave and could only be returned to the family after a year had passed. Therefore, a death certificate was glued to the burial shroud, usually on the cloth near the face, so that the body could be easily identified. Frale's reconstruction of the death certificate reads, "In the year 16 of the reign of the Emperor Tiberius Jesus the Nazarene, taken down in the early evening after having been condemned to death by a Roman judge because he was found guilty by a Hebrew authority, is hereby sent for burial with the obligation of being consigned to his family only after one full year." Dr. Frale noted that many of the letters were missing from the Shroud, and that Jesus, for example, was referred to as "(I)esou(s) Nnazarennos."[16]

In another interview, she said that "Those inscriptions are called 'traces of transferred writing', that is, traces of writing impressed on an object (in our case the Shroud) that has been in contact with a written text. The writing is in Hebrew, Greek, Latin and Aramaic. Thanks to computerised reading systems, those traces have been deciphered."[17]

The thesis of inscriptions on the Shroud has been criticized by both supporters and detractors of the authenticity of the Shroud. Among supporters, Ian Wilson and John P. Jackson have expressed their reservations about this approach. Jackson pointed out that "there is a long history of people finding things on the Shroud which are tied into subjectivity."[18]

Mark Guscin holds that "the whole affair is yet another example of things being seen on the Shroud in an attempt to come up with something new."[19]

Shroud skeptics are, well, skeptical: "Unusual sightings in the shroud are common and are often proved false, said Luigi Garlaschelli …. He said any theory about ink and metals would have to checked by analysis of the shroud itself."[20] According to Antonio Lombatti, "People work on grainy photos and think they see things. It's all the result of imagination and computer software."[21]

The critiques from the skeptics do not engage the proposal on a technical level by assessing the merits of the technology used in detecting the supposed inscriptions. But the critics have rightly drawn attention to the speculation and extrapolation underlying the reconstruction of the text. At best, the jury is still out on the claim of a "death certificate." But many great discoveries in science would not have been possible if researchers were not willing to chase tenuous clues and obscure data-sets. In the end, of course, the hypothesis has to be validated with compelling evidence.

2.6 Sudarium and Shroud – a Tale of Two Images

Another new dimension to research into the Shroud is its relation to the lesser-known Sudarium of Oveido. The Sudarium is believed to be the cloth that covered the face of Jesus after he was taken down from the cross. It is a 33 by 21 inch handkerchief-type linen cloth with bloodstains and serum stains. It is known to have been resident in Spain since at least 711 AD and is housed in the Cathedral of Oviedo.

It is possible that one of the burial cloths referenced in John's tomb narrative is the face-cloth. And, as indicated in the account of Lazarus' rising and contemporary excavations, Jewish burial customs call for a separate face-cloth. Once a person is deceased, Jewish custom required that the head be covered. It is believed to have covered the head of Jesus when he was brought down from the cross until he was wrapped in a shroud.

On the surface, the match between Shroud and Sudarium is overwhelming. The facial and neck stains on both cloths coincide with 70 matches on the front image and 120 on the back. The blood groups and the structure of the nose on both cloths match as well.

Research studies on the Sudarium began only in the latter part of the 20th century but the Spanish Center of Sindonology has since generated much useful data. A presentation from the Center gives an update on what has been learned:

> As a summary of the results obtained up to now from the forensic, geometrical and mathematical studies carried out by the EDICES, the following can be confidently stated

1. The Sudarium of Oviedo is a relic, which has been venerated in the cathedral of Oviedo for a very long time. It contains stains formed by human blood of the group AB.
2. The cloth is dirty, creased, torn and burnt in parts, stained and highly contaminated. It does not, however, show signs of fraudulent manipulation.
3. It seems to be a funeral cloth that was probably placed over the head of the corpse of an adult male of normal constitution.
4. The man whose face the Sudarium covered had a beard, moustache and long hair, tied up at the nape of his neck into a ponytail.
5. The man's mouth was closed, his nose was squashed and forced to the right by the pressure of holding the cloth to his face. Both these anatomical elements have been clearly identified on the sudarium of Oviedo.
6. The man was dead. The mechanism that formed the stains is incompatible with any kind of breathing movement.
7. At the bottom of the back of his head, there is a series of wounds produced in life by some sharp objects. These wounds had bled about an hour before the cloth was placed on top of them.
8. Just about the entire head, shoulders and at least part of the back of the man were covered in blood before being covered by this cloth. This is known because it is impossible to reproduce the stains in the hair, on the forehead and on top of the head with blood from a corpse. It can therefore be stated that the man was wounded before death with something that made his scalp bleed and produced wounds on his neck, shoulders and upper part of the back.
9. The man suffered a pulmonary oedema as a consequence of the terminal process.
10. The cloth was placed over the head starting from the back, held to the hair by sharp objects. From there it went round the left side of the head to the right cheek, where, for apparently unknown reasons it was folded over on itself, ending up folded like an accordion at the left cheek. It is possible that the cloth

was placed like this because the head formed an obstacle and so it was folded over on itself. On placing the cloth in this position, two stained areas can be anatomically observed – one over the "ponytail" and the other over the top of the back. ... The only position compatible with the formation of the stains on the Oviedo cloth is both arms outstretched above the head and the feet in such a position as to make breathing very difficult, i.e. a position totally compatible with crucifixion. We can say that the man was wounded first (blood on the head, shoulders and back) and then "crucified".

11. The body was then placed on the ground on its right side, with the arms in the same position, and the head still bent 20 degrees to the right, and at 115 degrees from the vertical position. The forehead was placed on a hard surface, and the body was left in this position for approximately one more hour.
12. The body was then moved, while somebody's left hand in various positions tried to stem the flow of liquid from the nose and mouth, pressing strongly against them. This movement could have taken about 5 minutes. The cloth was folded over itself all this time. The cloth was then straightened out and wrapped all round the head, like a hood, held on again by sharp objects. This allowed part of the cloth, folded like a cone, to fall over the back. With the head thus covered, the corpse was held up (partly) by a left fist. The cloth was then moved sideways over the face in this position. Thus, once the obstacle (which could have been the hair matted with blood or the head bent towards the right) had been removed, the cloth covered the entire head and the corpse was moved for the last time, face down on a closed left fist. This movement produced the large triangular stain, on whose surface the finger shaped stains can be seen and on the reverse side of the cloth, the curve inscribed on the cheek. Like the previous movement, this one could have taken 5 minutes at most.
13. Finally, on reaching the destination, the body was placed face up and for unknown reasons, the cloth was taken off the head.
14. Possibly myrrh and aloes were then sprinkled over the cloth.

The presentation also compares the marks on the Sudarium and the Shroud:

The part of the sudarium that was in direct contact with the face was the one called Reverse Left, so we should be able to show the existence of certain coincidences between the following anatomical elements of the face of the man of the Sudarium and that of the man on the Shroud:

1. The nose - a total area of 2,280 mm^2 on the sudarium and 2,000 mm^2 on the Shroud.
2. The superciliary ridges.
3. Absence of right cheekbone corresponding to the swelling observed in this area on the Shroud.
4. Swelling on the right side of the nose, c.100 mm^2 on the sudarium and c.90 mm^2 on the Shroud.
5. The tip of the nose, nostrils and nose-wings.
6. The position and size of the mouth, especially the blood flow on the right hand side, as already mentioned, first described by Ricci.
7. The chin.
8. The shape of the beard.

If we then observe the front left quadrant of the man of the Shroud and the front left quadrant of the sudarium of Oviedo we can see that there is an almost biunique correspondence between the position and size of the stains on each cloth...

If we observe the right frontal area, we can see that the drop of blood over the left eyebrow is geometrically compatible with the stain in exactly the same area on the sudarium. Both areas are 80 mm^2 and the relative position is practically the same on both cloths. It is interesting to note that this stain can be seen in two positions on the Oviedo cloth, a clear sign that the cloth was moved sideways over the face. The correspondence between the bloodstains on the two cloths is therefore acceptable (in the areas where the face is seen from the front) and the same can be said for the marks or traces visible on each. This correspondence is evident with the size of the stains, their relative positions on each cloth, and their origin. ...

Summing up the coincidences between the bloodstains visible on each face, we can say that it is possible to establish correspondence in the following aspects;

- The size of the stains is geometrically compatible, as is their relative position on each cloth.
- The blood is human, of the group AB on both cloths.
- The stains formed by blood shed in life are the same on each cloth.
- The stains occupy the positions predictable from the formation of the Shroud image ...

The bloodstains on the back of the man also correspond on the two cloths. On the Oviedo cloth they can be found in the lower left and right corners ...

From the forensic point of view, it is clear that the Shroud wrapped the dead body of a man who had been whipped, crowned with thorns and crucified. The sudarium of Oviedo wrapped the head of a body whose death is perfectly compatible with crucifixion and the wounds inflicted before death visible on the Shroud. The two deaths are very similar. The Shroud wrapped the whole body, including the head. The sudarium wrapped all the head of a corpse and touched the shoulders (especially the left one) and the back slightly. The correspondence between the bloodstains on the cloths is practically biunique i.e. each bloodstain has a corresponding stain on the other cloth as far as size and blood group is concerned, always bearing in mind (with the margin of error) that this correspondence is evident ALL ROUND THE HEAD. The stains on the left-hand side of the front show the sideways movement described by Jackson. The further the stains are from the middle of the face, the more this movement can be seen. It should also be said that the bloodstains on the head of man on the Shroud show signs of having been covered with another cloth. ...

We could ask the following question – what is the probability that two sets of stains formed at random, at different times by different bodies, could correspond to such an extent on a flat surface? Although we do not have the answer to this question yet, it is clear that the probability is very small. If we add to this the physical (time involved and formation mechanisms of the stains) and historical conditions, we are left with only one possible answer – everything seems to indicate that both cloths wrapped the same body and this body was that of Jesus of Nazareth, a Jew crucified in Jerusalem under the Roman governor Pontius Pilate ... in the place called Golgotha.[1]

Mark Guscin, who was cited earlier, is known for his work on the Sudarium as well as the Image of Edessa. Guscin points to the correspondence between Sudarium and Shroud as it relates to the crown of thorns:

The image of the back of the man on the Shroud is covered with wounds from the scourging he received before being crucified. The wounds on the man's back are obviously not reproduced on the sudarium, as this had no contact with it. However, there are thick bloodstains on the nape of the man's neck, showing the depth and extent of the wounds produced by the crown of thorns. This crown was probably not a circle, as traditional Christian art represents, but a kind of cap covering the whole head. ... The stains on the back of the man's neck on the Shroud correspond exactly to those on the sudarium.[2]

The same is true of stains from the beard:

Perhaps the most obvious fit when the stains on the sudarium are placed over the image of the face on the Shroud, is that of the beard; the match is perfect. This shows that the sudarium, possibly by being gently pressed onto the face, was also used to clean the blood and other fluids that had collected in the beard. Stain number 6 is also evident on all four faces of the sudarium. If stain 13 is placed over the nose of the image on the Shroud, stain 6 is seen to proceed from the right hand side of the man's mouth. This stain is hardly visible on the shroud, but its existence has been confirmed by Dr John Jackson, who is well known for his studies on the Shroud using the VP-8 image analyser. Using the VP-8 and photo-enhancements, Dr Jackson has shown that the same stain is present on the Shroud, and the shape of the stain coincides perfectly with the one on the sudarium. The gap between the blood coming out of the right hand side of the mouth and the stain on the beard is mapped as number 18. This gap closes as the stains get progressively more extensive on faces 1, 2, 3 and 4 while at the same time they are less intense. Stain number 12 corresponds to the eyebrows of the face on the Shroud. As with the beard, this facial hair would have retained blood and this would have produced the stains on the sudarium when it was placed on Jesus' face. There is also blood on the forehead, which forms stain number 10 on the sudarium. [3]

A 2015 scientific study titled "New coincidence between Shroud of Turin and Sudarium of Oviedo" compares the two cloths using X-ray fluourescence data:

The Sudarium of Oviedo and the Shroud of Turin are two relics attributed to Jesus Christ that show a series of amazing coincidences announced in the past. In this contribution, we describe the X-ray fluorescence analysis carried out on the Sudarium. Among the chemical elements detected, calcium shows a statistically significant higher presence in the areas with bloody stains. This fact allows correlating its distribution with the anatomical features of the corpse. A large excess of calcium is observed close to the tip of the nose. It is atypical to find soil dirt in this zone of the anatomy, but it is just the same zone where a particular presence of dust was found in the Shroud. The very low concentration of strontium traces in the Sudarium matches also well with the type of limestone characteristic of the rock of Calvary in Jerusalem. This new finding adds to others recently released and it strengthens the tradition that both cloths have wrapped the body of Jesus of Nazareth.[4]

At a 2015 conference, forensic specialist Alfonso Sánchez Hermosilla reported that

"From a forensic anthropological point of view and a forensic medicine point of view all the information that emerged from the scientific investigation is compatible with the theory that the Turin Shroud and the Sudarium covered the corpse of the same person. ... The "composition" of the textile structure of the Shroud and the Sudarium "is the same – substantially linen – the thickness of the fibres is identical, the fabric is hand spun with a "Z" twist, although they were woven differently: the Shroud has a herring-bone weave, while the Sudarium is taffeta. ... The morphological study of stains in both linens reveal an obvious similarity between them, ... not only in their relative position but also in their superficial size." Moreover, there is a "correspondence on the distances between the staining injuries which originated the stains".

Hermosilla confirms what others have said: some of the bloodstains appear to have been caused by the wounds traditionally associated with the Passion:

"The blood stains attributed to the thorns of the crown can be appreciated in both relics with a high similarity in the distance which separates them."

One of the stains on the Sudarium of Oviedo "seems to be compatible with some of the wounds inflicted by the Flagrum

Taxilatum" – the whip used to beat the man on the Shroud – "on the right hand side of the neck and also proves compatible with some of the imprints on the Turin Shroud, attributed to the same reason.

In the part of the Sudarium that "covered the right top area of the back there is a stain, located at the left bottom corner of the reverse of the linen, and could have been produced as a consequence of the orifice exit of the wound produced by the lance. ... Apart from this stain, there are some other indirect signs of the lance, such as plentiful clots of fibrin which appear in the called diffused stain and stain in accordion."

"All the information obtained from the studies and research" carried out on the Shroud of Turin and the Sudarium of Oviedo "is in tune with what one would expect - from a forensic medicine point of view - to happen to cloths with these characteristics were they to cover the head of a body featuring the kind of lesions Jesus of Nazareth suffered, just as the Gospels tell us."[5]

Finally, Juan Manuel Miñarro, a professor of sculpture, announced in 2016 that his fieldwork indicated that "the Shroud of Turin and the Sudarium of Oviedo 'almost certainly covered the cadaver of the same person.' This is the conclusion from an investigation that has compared the two relics using forensics and geometry." Miñarro's research

> found a number of correlations between the two relics that 'far exceeds the minimum number of proofs or significant points required by most judicial systems around the world to identify a person, which is between eight and 12, while our study has demonstrated more than 20.'
>
> Specifically, the research has discovered 'very important coincidences' in the principal morphological characteristics (type, size and distances of the markings), the number and

distribution of the blood stains, the unique markings from some of the wounds reflected on both of the cloths or the deformed surfaces.

There are 'points that demonstrate the compatibility between both cloths' in the area of the forehead, where there are remains of blood, as well as at the back of the nose, the right cheekbone and the chin, which 'present different wounds.'

Regarding the blood stains, Miñarro explained that the marks found on the two cloths have morphological differences, but that 'what seems unquestionable is that the sources, the points from which blood began to flow, correspond entirely.'

The variations could be explained by the fact that 'the contact with the [cloths] was different' in regard to duration, placement and intensity of the contact of the head with each of the cloths, as well as the 'elasticity of the weave of each linen.'[6]

2.7 The Shroud – a History

A Star is Born

"History is not a creed or a catechism, it gives lessons rather than rules; still no one can mistake its general teaching in this matter, whether he accept it or stumble at it. Bold outlines and broad masses of colour rise out of the records of the past. They may be dim, they may be incomplete; but they are definite."
John Henry Newman, *An Essay on the Development of Christian Doctrine*[1]

About the prior history of the Shroud of Turin, let us say this. Even if we find something that is known to be an ancient artifact, there is usually no accompanying narrative of its entire history. In fact, there is no logical requirement that we should know the history of the artifact before we can make a judgment about its authenticity.

We start with the fact of the artifact!

Let us study its characteristics and use these to explain the artifact the best way we can. The fact comes before theories about it and the theories – whether of origin or history – should always be tested against the (arti)fact. To thine own fact be true!

This means a judgment about the Shroud of Turin being the actual burial cloth of Jesus of Nazareth is not dependent on an exhaustive accounting of its entire history. We can make reasonable judgments about it simply from its knowable characteristics. This would be true even if we happened to first discover the Shroud in the 21st century.

Stephen J. Mattingly and Roy Abraham Varghese

Real-time History

Certainly, historical studies are valuable but inevitably the availability of historical data is limited by the very nature of history with its wars, famines, fires, tyrants, thieves and all manner of chaos and destruction. We are also limited by the human condition. No one lives for hundreds of years, let alone thousands, so we will have no living witness who can testify to the actuality of ancient events. And historians have their limitations. Herodotus, the "father of history," was also called the "father of lies!" Plutarch, another historian, titled one of his essays "On the Malice of Herodotus."

Mainstream historians today face one major impediment when it comes to the study of the Shroud: the radiocarbon dating verdict of a medieval origin. Since most of them apparently accept this verdict, they see no point in pursuing inquiries into the Shroud's pre-medieval history.

As has happened before, an influential error in one field has ripple effects on others. To give one extreme example, Trofim Lysenko, Josef Stalin's biology commissar, rejected the science of genetics because it arose from "bourgeois capitalism." In its place, he installed a version of scientifically discredited Lamarckianism as the ruling paradigm of Soviet science. This stricture had negative real-world consequences of various kinds including later crop failures. Fortunately, not all historians have accepted the flawed C14 conclusions. And veteran historians of the ancient world furnish additional relevant data. Valuable professional resources are thus available for a pre-medieval inquiry.

One dimension of the Shroud is especially striking in the context of history. Since we live in the era of You Tube, social media, 24/7 cable TV and special effect movies, we forget how recent is the ability to "capture" exact images of people and places. All the images we have of even the eras of the American and French Revolutions are paintings and sculptures, the products purely of artistic ability and media and not photonic mapping.

But with the Shroud, we have what is functionally a photograph: a one to one mapping of a person and in 3D! The Gospels give us a script that corresponds with what we see on the body and blood images. Microbiology tells us how it all happened. And other "witnesses," from the pollen and sand and the cloth design to the Sudarium of Oviedo, take us to first century Palestine.

Smoking Gun

We have pointed out in the introduction that a study of the Shroud's history needs to be driven by a focus on its unique nature. Only in the light of the Shroud as it is today can we study its historical antecedents, especially the pre-medieval years.

Viewed from this perspective we see why the quest has to involve not just historians but scientists from various disciplines. And the very nature of the artifact affects the way in which we study its history. It is, quite simply, a smoking gun – a gun that literally leads back to a body! Normally, we find the body first and then seek the smoking gun that identifies the perpetrator. In this instance, however, we are confronted with a smoking gun that bears witness to a body.

All sensible Shroud studies focus on the smoking gun and consider the body in its light. Shroud skeptics, on the other hand, completely miss this point and leave behind both gun and body preferring theory over fact, speculation over substance. For instance, those who dismiss the Shroud as a medieval painting ignore the various problems afflicting this view including the question of how a medieval artist could have created a work of art with the unique properties of the Shroud. Despite such ongoing obtuseness, independent third parties have recognized the sheer volume of circumstantial evidence testifying to the Shroud's presence over two millennia.

As previously noted, in investigating the Shroud's history and prehistory (by which we mean the pre-medieval history), we realize that our judgment on the artifact is not dependent on our ascertaining every detail of its history. But there seems to be enough of a body of fact available for us to make plausible deductions and reach reasonable conclusions about how the Shroud appeared in fourteenth century France. We have an assortment of clues from which we can paint a plausible picture. While we cannot expect a scientific theory validated by replicable experiment, we do have enough evidence to show that, through the centuries, there have been reports of an artifact matching much of what we know about the Shroud. In short, we are not dealing with a black hole when it comes to the Shroud's pre-1357 history.

The extraordinary hard fact that we have in our hands today – an artifact that could only be fully understood in the light of modern science – has left a trail of "soft" facts leading back to the first century.

Following the Star

Think of it this way. The stars we think we see today are actually glimpses back into time: since their light has to travel vast distances to reach us, what we are seeing before us is how some of them looked millions of years ago. With the Shroud we are seeing the star as it is here and now. Our deductions of its history are simply a matter of examining the witnesses to its starlight in the past. The scientific investigation of the Shroud is a study of the star itself while an inquiry into its history is a study of its starlight. What is more, it glistens even more brightly today than at the time of its inception (for instance, its "negative" attribute was discovered only after the invention of photography).

Knowing that a search into the Shroud's past cannot go beyond an inquiry into traces of its starlight, we should modulate our expectations, claims and critiques accordingly. Let us press the analogy further.

Astronomers estimate the number of stars in the Universe not on the basis of the twinkling objects they are able to see. They apply a variety of methods for their calculations of the stellar population: using our own galaxy as a model from which to make estimates of the size of other galaxies; determining the underlying aggregate of bright and not-so bright stars from observations of a few distant bright objects; deducing the number of supernovae that had to collapse to form today's heavy elements; using galactic mass to infer the volume of stars. In short, this is a work of extrapolation and analogical thinking – much like the study of the Shroud's starlight.

Some academics scorn and sneer at attempts to inquire into the pre-medieval history of the Shroud. If their approach were adopted in other realms of exploration from astronomy to the search for a cancer cure, we might as well call a halt to any inquiry that goes beyond what is immediately observable. Those who look down their nose at this kind of exploration choose not to look beyond their nose. But there is no reason why we should join them in their self-captivity. Evidence is what matters in the end but, as the astronomers can tell us, the evidence for a star is not always twinkling "like a diamond in the sky."

From Edessa to Constantinople

The Shroud's starlight begins with scattered reports about Jesus' burial shroud in the first centuries of Christianity (although the Discipline of the Secret precluded Christians from publicly speaking about sacred matters); followed by the sudden prominence of an image of Jesus "not made with human hands" (Edessa 400 A.D.); then historical reports of a full-length body image of Jesus matching the description of the Shroud of Turin beginning in Edessa and reaching a climax in tenth century Constantinople; and finally the emergence of the present-day Shroud in France after the sack of Constantinople by French and other European Crusaders.

The Image of Edessa and the Constantinople full-body image each had a dramatic impact on popular depictions of Christ: starting in the sixth century, icons matched the face of the Man of the Shroud while the transfer of the Image of Edessa to Constantinople was accompanied by Lamentation art (the laying out of Jesus' body on a slab). We may not have thoroughgoing documentation of the Shroud from the first century to the present but an argument from silence against it would silence much of what we call history. This argument is flawed in any case as historians themselves have testified: "Another difficulty with argument from silence is that historians cannot assume that an observer of a particular fact would have automatically recorded that fact. Authors observe all kinds of events but only record those that seem important to them."[2]

Three Startling Signposts

What is unique and utterly inexplicable in the prehistory period was the emergence and conjunction of three unprecedented phenomena:

- A focus on the shroud left behind in Christ's tomb in the Church Fathers and other early Church writings along with references to the continued existence of the shroud: never before has a burial cloth played this kind of role or any kind of role in secular or religious history
- The claim to possess an image of Christ produced without "human hands" that was a commonplace by the fourth to sixth centuries
- The undeniable presence in Constantinople, the reliquary of Christendom, from the tenth to the thirteenth centuries of a rarely displayed shroud of Christ that appeared to bear the features of the Shroud of Turin

How did these three ideas, unique in human history, arise? The fact of these phenomena is evident in historical documents. This is the best that conventional history can provide.

Historians as historians cannot be expected to draw any further conclusions from these three signposts. The onus is on the observer to be a "synthesizer" who can rationally follow the path to which they point and see the connection between starlight and star.

The Takeaway

Thomas de Wesselow hones in on the historical takeaway from all this with pellucid clarity:

> Being confronted with genuinely ancient objects of unknown provenance is a common experience for the museum curator...The "lost" 1300 years and the image origin may always remain unexplained...but data sufficient for authentication have been obtained from other aspects of the Shroud. The dating, geographical origin, and association with

Christ are indicated not by an isolated feature or datum, but by a web of intricate, corroborating detail as specific as that used in the authentication of a manuscript or painting and certainly as reliable as many other archaeological/historical identifications which are generally accepted.

Consider, for example, the case of one of the most celebrated artifacts in the British Museum, the Portland Vase. This remarkable piece of glasswork, exhibiting a refined cameo technique invented by the Romans in the first century BC, is first documented in 1600-1, when it was in the collection of an Italian cardinal. No one knows where the vase came from, but scholars are sure, none the less, that it is a genuine Roman treasure, based purely on their study of the object. The 1,600-year gap in its history does not make it a Renaissance forgery. Similarly, the Egyptian shroud in the Metropolitan Museum of Art ... turned up out of the blue in the hands of a Cairo antiquities dealer in the early twentieth century, but no one doubts that it is an authentic, second-century artifact. If the object is convincing in itself, lack of provenance means nothing.

The Shroud of Turin can be traced back 1,000 years, at least. That earns it the right to be judged by the same sort of criteria as are used to authenticate the Portland Vase and the Metropolitan Museum shroud. Like them, it stands as "testimony to its own authenticity." Its early history is a separate matter entirely.[3]

We can, then, justifiably apply Newman's statement to the Shroud's prehistory: "Bold outlines and broad masses of colour rise out of the records of the past. They may be dim, they may be incomplete; but they are definite."

The starlight of the past bears witness to the star of the present.

Three's a Shroud

The Shroud's history can be divided into three phases. To start with, its history can be divided into prehistory (before 1357) and "public" history (after 1357). But then the "prehistoric" period can be further divided into the first through the fourth centuries and the fifth through the thirteenth centuries. Thus, we get the three phases: Phase One, first through the fourth centuries; Phase Two, fifth through the thirteenth; and Phase Three, the fourteenth and after: ancient, Byzantine, medieval/modern: root, shoot and fruit.

Public History

There is not much dispute about the Shroud's public history which at a high level runs something like this:

1349-1357 – Geoffrey DeCharny I, a French knight, writes Pope Clement VI to tell him that he is going to build a church in Lirey, France in gratitude for favors received. The Shroud is shown to the public at this church somewhere between 1353 and 1357 in public exhibitions. The local bishop attempts to prohibit these exhibitions.

1389-1390 – King Charles VI of France tries to gain possession of the Shroud which is now owned by DeCharny's son, Geoffrey II. Bishop Pierre D'Arcis asks Anti-Pope Clement VII to prohibit exhibitions of the Shroud but his request is denied.

1453-1463 – Margaret de Charny, Geoffrey II's daughter, turns the Shroud over to the Duke of Savoy as part of a financial transaction.

1473 – The Shroud is moved for the first time to Turin, a part of the Savoy realm. The Shroud travels with the Savoy family across various parts of Europe.

1518-1535 – The Shroud is exhibited in Chambery from 1518. A fire in the church there melts the reliquary in which the Shroud is kept and causes damage to the Shroud as well. Nuns stitch a backcloth for the Shroud.

1578 – The Shroud takes up residence in Turin (thus escaping the French Revolution's orgy of destruction).

Pre-History

It is when we turn to the Shroud's "prehistory" that we encounter conroversy. This prehistory is denied by those who believe that the Shroud is the creation of a medieval artisan and/or those who accept the verdict reached by the radiocarbon dating team. But the radiocarbon dating results, we have seen, can tell us nothing about the age of the Shroud because the samples were not decontaminated at a microbial level prior to the testing. And, as we will continue to show, the claim that the Shroud is a medieval creation fails to address any one of the Shroud's unique attributes. The Shroud in its three-dimensional photographic splendor is a hard fact that cannot be washed away by fact-free speculation and erudite esoterica. With a continued focus on facts, we will pinpoint what is known about any artifact bearing the Shroud ID prior to the 14th century exhibition in France.

Given the controversy and confusion, our concern here is primarily with Phases One (first through the fourth centuries) and Two (fifth through the thirteenth centuries). Granted, we know very little about the Phase One period beyond fragmentary reports given the challenges described below. But Phase Two is much better documented and Phase Three is a part of the recorded history of Europe. Despite the unevenness of the available data, we know enough to recognize the plausibility of a path connecting the various phases as it proceeds from an unmarked trail with twists and turns to a bumpy country road to a high-speed highway.

Several factors stood in the way of insights into the Shroud in its earliest days:

- the persecution and execution of Christians in the first three centuries of the church;
- the Discipline of the Secret that required Christians to refrain from talking about the sacred mysteries at the heart of their faith;
- the fact that Jews considered burial cloths to be unclean; and, much later,
- the iconoclastic controversies that resulted in the destruction of many relevant records.

Shooting Stars

The Shroud, it is claimed, is the cloth in which the body of Jesus was wrapped after he was crucified and laid to rest in a tomb. But is there is any indication in early Christianity of the continued existence of this shroud? We know there are documents speaking of an imprint of Jesus not made by human hands from the fourth/fifth centuries and the presence in Constantinople of his burial linens from the tenth century. But are there any earlier data sources?

This is the summary answer:

- Second century Gnostic gospels state "that Jesus' shroud was removed from the tomb and saved."
- There are accounts of the preservation of the shroud from Nino (296-338) and other early visitors to Jerusalem.
- A liturgical text from the seventh century speaks of Peter and John seeing the recent imprint of Jesus on his burial cloth.
- John of Jerusalem, secretary of the Patriarch of Antioch, writes in 764 A.D. of an image made by Jesus of himself that still exists.
- A Good Friday sermon by Pope Stephen III in 769 A.D. speaks of a transformational image of the whole body of Jesus imprinted on a cloth.

- A text from the eighth century speaks of a first century tradition of a sacred image of Jesus being moved from Judaea to Syria when Titus and Vespasian sacked Jerusalem.
- From the third through the ninth centuries, numerous Fathers of the Church and other leaders preach about the presence of the shroud in the empty tomb.

Second Century

The significance of the references in the Gnostic gospels is highlighted by historian Dan Scavone:

In the second century (about 100-200 A.D.), several accounts were written about the life of Christ. These biographies are similar to the Gospel accounts in the Bible. For various reasons the early Church Fathers did not include them among the 'official' texts of the Bible. Some of these writings contain incorrect religious teachings; some are just copies of the Gospels with a few additions. Hence we have called them 'unofficial.' The usual word for these books is 'apocryphal' or 'hidden' books. But because they were excluded from the Bible does not mean that they are utterly false. They agree with the Gospels on many points. As books actually written in the second century, they are valuable source materials for that time. Most importantly, these texts say that Jesus' shroud was removed from the tomb and saved. Writers of the second century, therefore, knew of the existence of this sheet in their own day. The first of these apocryphal books is called the *Gospel of the Hebrews*. The author is anonymous (unknown) as is the case with all these apocryphal books. We have only fragments from it, for most of it has been lost over the centuries. One key surviving passage says, 'After the Lord gave his shroud to the servant of the priest [or of Peter; the actual word is not clear], he appeared to James.' The *Acts of Pilate* is another apocryphal book of the second century. It states that Pilate and his wife preserved the shroud of Jesus. It suggests that they were sorry for their part in his death and were now Christians. These two books, along with the *Gospel of Peter*, the *Acts of Nicodemus* and the *Gospel of Gamaliel*, show us that second century writers knew about the Shroud in their day. They disagree about who saved it from the tomb, but they agree that it had been saved. The silence of the "official" Biblical stories about the preservation of the shroud is countered by these books.[4]

Fourth to Seventh Centuries

Scavone also considers the significance of what is known from Nino and early first millennium visitors to Jerusalem:

The Shroud record is again silent for nearly two centuries. These are centuries of persecution of Christians. The earliest martyrs died for their faith during this period. The Shroud may have continued to be hidden away for its own protection. The next reference to it comes in the biography of a young girl named St. Nino. She had visited Jerusalem during the time of Constantine. Constantine (312-337 A.D.) was the first Christian to rule the Roman Empire. It was he who put an end to the religious persecution of Christians. He also decreed that death by crucifixion should be outlawed. St. Nino took a great interest in the relics of Jesus' Passion (the sad events from the Last Supper on Thursday, through Good Friday, to Easter Sunday). These relics included ... his burial sheet. Jesus' shroud, she reported, had been preserved by the wife of Pilate, who then gave it to St. Luke who hid it away. After some time, St. Peter found it and kept it. St. Nino's account is proof that in fourth-century Jerusalem people still knew of the Shroud's existence. ....

After St. Nino ... There is still more evidence for the Shroud of Jesus in Jerusalem ... 1) Around the year 570 a pilgrim to the Holy Land, Antonius of Placentia, wrote of seeing a cave on the banks of the Jordan River. In it were seven cells, or rooms.

In one of the cells was found 'the sudarium which was upon Jesus' head.' 2) Not much later, St. Braulion of Saragossa, Spain (585-651) also saw in Jerusalem the 'linens and sudarium in which the Lord's body was wrapped.' He adds something which might be good to keep in mind: 'There are events of which the Gospels do not speak ... such as preserving the burial sheet.' 3) Next comes the wording to the 'Mozarabic Liturgy.' ... This text was originally written in the sixth century, so it is contemporary with Antonius and Braulion. The lines which intrigue the student of the Shroud read, 'Peter ran with John to the sepulcher. He saw the linens and on them the recent *traces* of the death and resurrection.' Could this be the first hint that the surviving grave wrapping showed an image? ... The historical records placing the Shroud in Jerusalem are not very persuasive. They may refer to some cloth other than the real burial sheet of Jesus. However, they cannot be discounted completely, especially Arculf's story. They do represent part of the Shroud mystery.[5]

Keeping a Secret

None of this constitutes an argument for the preservation of today's Shroud from the time of the burial of Jesus. At best, they are flickering traces of starlight from a long time ago in a galaxy far far away. And there are any number of reasons why we cannot expect anything more. For instance:

- Christians were fighting for survival in the first three centuries of the church and it is hard to imagine that they would have the ability or the means to set down every detail of their actions and priorities at a time when martyrdom was the "new normal."
- Until the fifth century, Christians were instructed to refrain from discussing their sacred mysteries for fear of profanation and ridicule: this was the Discipline of the Secret described further here.

- Burial linens were considered "unclean" by the Jews and any mention of the continued existence of Jesus' shroud would have been considered scandalous.

An early researcher on the Shroud, Theodora Cogswell, gives us a sense of perspective on the need for discretion in those turbulent times:

> Can we wonder that the Apostles and their companions anxiously hid away from the world at large this record of their Lord? Is it strange that they made no open mention of it in the widely circulated Gospels and Epistles which were sure to fall under hostile eyes?... Had the Shroud been openly mentioned in the Acts or Epistles as if it were still existing, undoubtedly the Roman authorities would have instituted a determined hunt for it... References to the Shroud have been overlooked by translators who were not on the alert for such material.[6]

Note too the Discipline of the Secret was in effect from the first to the fifth centuries. Those who had been initiated into the faith were urged to hold in confidence its mysteries. In *Concerning those who are initiated into the Mysteries* (380 AD), Ambrose of Milan, for instance, declaimed that "the mystery ought to remain sealed up with you ...that it be not made known to thou, for whom it is not fitting, nor by garrulous talkativeness it be spread abroad among unbelievers."[7] Said Basil the Great, "the awful dignity of the mysteries is best preserved by silence."[8]

With all this in mind, let us consider some of the available texts.

The Gnostic gospels

"And when the Lord had given the linen cloth [shroud] to the servant of the priest, he went to James and appeared to him." Second century *Gospel of the Hebrews*[9]

"To identify himself, the Lord shows Joseph of Arimathea the Shroud and the sudarium." Second century *Mysteries of the Acts of the Saviour*[10]

"The image of the king of kings was embroidered and depicted in full all over it [the glorious robe]." Third century *Hymn of the Pearl*[11]

"The *Acts of Pilate* show him [Jesus] with Joseph of Arimathea, saying, 'I am Jesus ... you wrapped me in a clean shroud (*sindone munda*) and you put a cloth (*sudarium*) on my face', before showing him where they lay."[12]

Nino

"And they found the linen early in Christ's tomb ... When they found it, Pilates's wife asked for the linen, and went away quickly to her home in Pontus, and she became a believer in Christ. Some time afterwards, the linen came into the hands of Luke the Evangelist, who put it in a place known only to himself."[13]

"St Nino, the fourth century apostle of Georgia, [says] the linen cloths passed via Pilate's wife to St Luke. The shroud, however, 'is said to have been found by Peter, who took it and kept it, but we know not if it was ever discovered.'

"This Petrine tradition must have been a persistent one because about 850 the Syrian, Ishodad of Merv, firmly believes it. He says that the burial linens were given to Joseph, the Lord of the grave, but 'the shroud (*sudara*, Syriac for *sudarium*) Simon took, and it remained with him. And whenever he made an ordination, he arranged it on his head and many and frequent helps flowed from it, just as even now leaders and bishops of the Church arrange their turbans that are on their heads and about their necks in place of that shroud.'"[14]

Holy Saturday liturgy of the seventh century Mozarabic Rite

"Peter ran with John to the tomb and saw the recent imprints (*vestigia*) of the dead and risen man on the linens."[15]

Eighth Century Writings

John of Jerusalem, secretary of Theodore, Patriarch of Antioch (764):

"Actually Christ Himself made an image, the one that is told not made by human hands, and until today it still exists."[16]

Pope Stephen's Good Friday Sermon (789):

"He stretched His whole body on a cloth, white as snow, on which the glorious image of the Lord's face and the length of His whole body was so divinely transformed that it was sufficient for those who could not see the Lord bodily in the flesh to see the transfiguration made on the cloth."[17]

The phrases "stretched His whole body on a cloth, white as snow" and "and the length of His whole body" are believed to be later although pre-944 interpolations. But even as such they bear witness to a common belief in the presence of an image of Jesus' whole body.

Next, we have an intriguing but similarly tenuous account of the transport of an image of Jesus from first century Jerusalem to Syria. It was presented in the fourth session of the Second Council of Nicaea (787 A.D.) and attributed to the Church Father Athanasius (328-373):

But two years before Titus and Vespasian sacked the city, the faithful and disciples of Christ were warned by the Holy Spirit to depart from the city and go to the kingdom of King Agrippa, because at that time Agrippa was a Roman ally. Leaving the city, they went to his regions and carried everything relating to our faith. At that time even the icon with certain other ecclesiastical objects were moved and they today still remain in Syria. I possess this information as handed down to me from my migrating parents and by hereditary right. *It is plain and certain why the icon of our holy Lord and Savior came from Judaea to Syria.*[18]

The eminent patrologist Brian Daley S.J. informed us that this is not today considered an authentic work of Athanasius. Hence it is described as Pseudo-Athanasius and probably originated from Syria or Palestine in the eighth-century. Like the Gnostic gospels, it simply serves as testimony to a tradition or legend, in this case of an image of Jesus being taken from Israel to Syria.

Heralds of the Shroud

Finally, we have the numerous references to the finding of the shroud in the empty tomb in the teaching of Christian leaders of every century. Of course, these are not accounts of the preservation of the shroud with the image. Their importance lies in the foundational importance given to the shroud of Jesus. This is a unique occurrence in the history of religion and history itself. A burial cloth is an integral part of the package in the new religion. A full list of those who have preached on the shroud is found in Pietro Savio's Sindonological Prospectus.[19] The preachers range from Origen in the third century to Hilary of Poitiers, Ephrem of Syria, Cyril of Jerusalem and Ambrose in the fourth to Epiphanius, John Chrysostom, Augustine, Cyril of Alexandria and Leo the Great in the fifth. The leaving behind of the burial cloth was an announcement of Jesus' rising from the dead. Hence its importance.

We have seen that those who accept the authenticity of the Shroud include non-Christians and non-theists. Hugh Schonfield, well known for rejecting traditional Christianity, held that the Shroud was genuine and that it was known to the first Christians. He asks, "whether at last there has been restored to us through the agency of science that most wished for object of Christian faith – to see the face of Jesus" and says, "It cannot be too strongly stressed that the Holy Shroud is in quite a different category from other objects of devotion. The discovery was made possible by the advance of science, which so often in seeming contradiction to religious belief, has as often confirmed it by what it has brought to light."[20] He then goes on to say that "Outside the Bible the oldest reference (to the Shroud) is contained in the Gospel of the Hebrews, which dates at the latest from the beginning of the second century, and by many of the Church Fathers was held to be the original Hebrew of St Matthew's Gospel. St Jerome, in a quotation from it which is tantalisingly short, introduces the following: 'Now the Lord, when he had given the shroud to the servant of the priest, went to James and appeared to him.'" ... Referring also to Nino, he says, "Taking the above references together, it is evident that, hazy as the tradition is, a case can be made out for the possession of the Shroud by Peter, the chief of the Apostles." He also cites the challenges faced by the first century Christians: "What became of the Shroud after his death? It was still missing early in the fourth century, but the Roman wars in Palestine which devastated the country and persecution of the Christians would no doubt lead to the secreting of the relic and to the loss soon afterwards of all knowledge of its hiding place."[21]

Those to whom all this is given, might expect much more. But they should tailor their expectations to the context. Here a word of counsel from archaeologist William Meacham for those with "great expectations" may be helpful:

> "The looting of Edessa (Urfa, Turkey) by 12th century Turkish Moslems ... yielded 'many treasures hidden in secret places, foundations, roofs from the earliest times of the fathers and elders...of which the citizens knew nothing' (Segal 1970:253). Similarly, it was not uncommon for manuscripts, works of art, and relics kept in monasteries gradually to drift out of the collective memory; the most notable example is the Codex Sinaiticus, which reposed in a Sanai monastery for over 1,000 years, its importance totally unknown to its keepers.... Being confronted with genuinely ancient objects of unknown provenance is a common experience for the museum curator."[22]

Apocryphal gospels, pilgrim narratives, liturgical texts, homilies – this is what we have in the first four centuries. We might call them shooting stars. But when we come to the fifth and sixth centuries we have a Big Bang – an explosion of artwork that bears a family resemblance to the face on the Shroud.

Big Bang

The adage "one picture is worth a thousand words" applies as much to historical reconstruction as to communication. Hence its relevance in any inquiry into the Shroud's history. The Shroud, it has been argued, is the image that launched a thousand icons. Not medieval icons but icons from 550 A.D. onwards.

The sixth century "Big Bang" in artistic representations of Jesus has been traced to the re-discovery of a long missing image of Jesus, the Image of Edessa (later called the Mandylion). Another artistic revolution that resulted in Lamentation Art, depictions of the laying out of Jesus' pierced and battered body, coincided with the transfer of this same image to Constantinople (tenth century). Both kinds of representation resulted in a sea change in the manner in which Jesus was subsequently portrayed. And both were, if we might use the expression, spitting images of the Man of the Shroud.

The earliest depictions of Jesus took a page from Greco-Roman art and portrayed him as a beardless youth. But right after the reported re-discovery of the Image of Edessa in the fifth century, Jesus was portrayed with a beard and shoulder-length hair and Semitic features missing in the early depctions. There were other features of this portrayal – of an eyebrow, the nose, a pattern on the forehead – that were specific to the Shroud image. The most famous representations from the sixth century are the mosaics at Ravenna and the Christ Pantocrator icon at the Sinai monastery of St. Catherine.

Emanuela Marinelli points out that "The similarity between the Shroud face and most of the depictions of Christ known in art, both Eastern and Western, is clear and cannot be attributed to pure chance; it must be the result of a dependency, mediated or immediate, of an image from … a common source. We can identify several elements on the Shroud that are not regular, hardly attributable to the imagination of the artists, that make us understand how the ancient representations of Christ's face depend on the venerated relic. It is reasonable to think that in the early days of the Church, the Shroud has been kept hidden for various reasons. During this period, for the representation of Christ they only used symbols or they applied to the figure of Christ appearances derived from other religions. After the victory of Christianity, sanctioned by Constantine in 313 with the Edict of Milan, a new image of the face of Jesus began to spread, which is characterized by not too long beard, mustache, narrow, tall and stately face, with long hair, falling on His shoulders, and sometimes with a middle line that divides them."[23]

The pioneer in exploring the cause and effect relationship between the Man of the Shroud and the Jesus icons was Paul Vignon, the early 20th century French researcher referenced before.

Vignon discovered fifteen random features of the Shroud image, some of which were caused by the material of the Shroud itself, that were found in later icons. These features, later called Vignon marks, were categorized under fifteen heads by Ian Wilson.

- Starkly geometric topless square (3-sided) visible beween the eyebrows on the Shroud image
- V-shape visible at the bridge of the nose
- A transverse streak across forehead;
- A second V-shape inside the topless square
- A raised right eye-brow
- An accentuated left cheek
- An accentuated right cheek
- An enlarged left nostril
- An accentuated line between enlarged between the nose and the upper lip
- A heavy line under the lower lip
- A hairless area between the lip and the beard
- The fork in the beard
- A transverse line across the throat
- The heavily accentuated "owlish eyes"
- Two loose strands of hair falling from the apex of the forehead[24]

It is possible that the commonality of random features may be coincidental in two or three cases but coincidences between fifteen such features are simply too many to be the product of chance. Skeptics have taken the tack of asking: which came first, the chicken or the egg? But given the "without human hands" fame of the Image of Edessa it is unlikely to have been a copy of an existing icon. And the almost universal acceptance of the particular portrayal found on the icons indicates it had some sort of special authority. Moreover, some of the markings are the result of creases in the Shroud and it makes no sense to say that an icon painter would have deliberately created an "original" with these flaws.

This train of thought is well developed by Maurus Green:

> Vignon has highlighted evidence of another kind – the hundreds of icons and *Mandylions* that strongly indicate the presence of the Shroud in the East from the sixth century. This is the earliest we can expect to hear of it in view of a long sequence of events and universal attitudes hostile to its disclosure and compelling its guardians to keep it a close secret: the Jewish horror of "impure" burial linens combined with the Jewish and Roman persecutions; the Christian shrinking from crucifixion and its detailed portrayal in art, an attitude that lasted many centuries; and the continuous quarrels about sacred images, both affecting and affected by the Christological controversies, from the earliest times to the final defeat of Iconoclasm.
>
> The late B. G. Sandhurst called these images of Christ "the Silent Witnesses" which steadfastly direct our attention to the Shroud. How do they do this? They point silently with the strange anomalies or disfigurements with which their artists felt compelled to adorn them. All these anomalies are to be found on the Turin Shroud (Plate 13), where they were produced either by the wounds and bruises of the Man of the Shroud or by faults in the linen accentuated by the stains of the imprint. ...
>
> Probably a very few artists actually saw the death mask of the Shroud but they seem to have reproduced its anomalies and

mistakes so faithfully that subsequent artists felt bound to copy them. A careful study of the characteristics and anomalies common to the *Mandylions* (Plates 7- 11), Byzantine Christs (Plates 1-6) and the Shroud, (Plate 13), will reveal what Vignon, Wuenschel and Sandhurst mean when they say that the Turin Shroud is the prototype of the Byzantine Christ and indeed the more remote origin of his traditional likeness in every school of art down to the present day.

A most striking confirmation of this theory can be experienced by the reader. Let him show a *positive* photograph of the Face of the Shroud to someone who has never seen it nor heard of the Shroud, and ask him whose image it is. He will get only one answer. The only explanation I can see for this recurrent phenomenon is that the ancient artists who copied the negative of the Shroud and gave us our traditional Christ, did their job so well that when the camera revealed the secret of its mysterious mask the resemblance was obvious. They did, up to a point, transpose negative details, e.g. the nose, so dark in the Shroud image, becomes of natural tone in the pictures. Other points, however, were not recognised, e.g. the dark-coloured closed eyelids are copied as wide open eyes; the drawing of the mouth is badly affected by the lack of understanding just where the lights and darks are inverted in the Shroud image.

A special feature of this iconographic evidence is the evident likeness of the isolated head of the Shroud to the Mandylions. With long hair, staring eyes and absence of neck, it almost seems to be their negative. Could this similarity, coupled with the anomalies common to both, give us the moral certainty that the Shroud was the unique *acheiropoietos*, kept, as Vignon believed, in some monastery easily accessible to the theologians and artists of Edessa? [25]

Marinelli takes up the theme: "Starting from the 6th century also in the East spreads a particular type of portrait of Jesus inspired by the

Shroud: it is the majestic Christ, with a beard and mustache, called the Pantocrator (Almighty), of which there are splendid examples in Cappadocia. There is an evident inspiration from the Shroud in the face of Christ in the silver vase of the 6th century found in Homs, Syria, now in the Louvre in Paris, and in that of the silver reliquary of 550, from Chersonesus in the Crimea, which is in the Hermitage in St. Petersburg."[26]

Duke Medical Center doctor Alan Whanger developed a method to determine points of congruence between various images. In the peer-reviewed publication *Applied Optics*, he published a paper on his image analysis technique, called Polarized Image Overlay Technique (PIOT). Under the PIOT system, "The two images for study are projected onto a suitable screen through polarizing filters, and these superimposed images are then viewed through a third polarizing filter. This procedure allows a dynamic comparison of both large features and fine details."[27]

Applying PIOT, Whanger found that "Comparing the [sixth century] Christ Pantocrator to the Shroud of Turin reveals at least 170 points of congruence, making this the most accurate non-photographic depiction of the Shroud that we have found (we have studied hundreds of icons)." Further, "Our comparisons over the years of hundreds of icons with each other and with the Shroud face have demonstrated that the model was the Shroud of Turin. Of course, there are differences in interpretation. Some artists are more skillful than others. ... Some of the artistic productions, of course, are more accurate than others. Interestingly, where they disagree with each other, they very often agree with the Shroud."[28]

As will be seen, the full-body image of Jesus on the Shroud came into focus especially in Constantinople. This had just as important an impact on the art of the day (the Byzantine *epitaphios* image of the corpse of Jesus in particular). One of the greatest contemporary Byzantine historians, Hans Belting, has highlighted this connection: "There was what was believed to be the authentic relic of the Holy Shroud preserved in the chapel of the Palace before it ended up in Turin. The existence of the true likeness of the buried Christ justified the creation of our [*epitaphios*] icon; with time, the icon came to reflect a shift of emphasis to the burial proper, which explains the burial position of the crossed hands."[29]

That there was a Big Bang cannot be denied. But what lay behind it? We have suggested that it was the Shroud. Some have said that the Shroud is the Artifact Formerly Known as the Image of Edessa. So to Edessa we shall go.

The Artifact Formerly Known as the Image of Edessa

The sixth century Big Bang event that transformed subsequent Christian art was the sudden pre-eminence of an image of Jesus reputedly "not made by human hands." Traditions associated with the image speak of it having been hidden away from enemies and being rediscovered centuries later (this re-discovery, we have said, apparently triggered off the new artistic depictions). Mainstream historians, in fact, agree that awareness of a uniquely originated image became prominent between the fifth and sixth centuries. These historians include the Jewish scholar Ernst Kitzinger of Harvard, an authority on Late Antique and Byzantine art; Glanville Downey, the author of the definitive history of ancient Antioch; and Mark Guscin, cited earlier, the author of the only two scholarly books published on the Image of Edessa.

Kitzinger observed that that the Image of Edessa was proclaimed as an image "believed to have been made by hands other than those of ordinary mortals"[30] Downey notes that the Jewish quarter of Antioch called the Kerateion "was regarded as having special religious associations. It possessed an image of Christ ... which was an object of particular veneration."[31] About the Image of Edessa, Guscin writes, "The image first appears in the Syriac work known as the *Doctrine of Addai*, which in its present form would appear to date from about AD 400."[32] It should be noted that copies were made of the original image. (The *Chronicle* of Michael the Syrian, for instance, recounts an account of a monophysite merchant making a copy of the Image of Edessa.[33])

Edessa is important because of its connection to ancient Christianity. Most people are unaware that Christianity began and flourished first in Asia. After Jerusalem, the major centers of the early church were Antioch and Edessa in Syria and Alexandria in Egypt. Peter was in Antioch before Rome and his disciple Mark in Alexandria. Hence Antioch, Alexandria and Rome are called the Petrine Triad. Antioch is where the followers of Jesus were first called Christians and it was from here that the Apostle Paul set out on his missionary journeys. Edessa was the first kingdom whose ruler (named Abgar, one of several with the same name) became a Christian.

The circumstances of the Edessa king's conversion are lost in legend but clearly something dramatic had to happen for such a momentous change. Later accounts talk of the king receiving a letter from Jesus and/or a miraculous imprint of his image. In his magisterial work on the Image of Edessa, Mark Guscin writes that

> The Image itself and the related texts ... cover a period of almost two thousand years, from the textual origins in Eusebius (and the legendary origins dating back to the life of Christ in Jerusalem) to the most recent painted icons. Edessa was in fact the first kingdom in the world to adopt Christianity as its official religion (most probably ca.200 AD), and both the Image and the supposed letter from Abgar to Christ and the latter's reply are major components in the argument to establish the early arrival of the new religion in the area."[34]

With respect to its later history, Guscin observes that "The Image was kept in Edessa even when the city was lost to the Byzantine Empire and was thus conveniently far removed from the iconoclastic crisis. Towards the middle of the tenth century it was finally taken to Constantinople. After a ceremonious arrival, it was kept in the Boucoleon and, apart from making an appearance in some pilgrims' lists of relics they had seen, is hardly mentioned again. After the sack of the capital during the Fourth Crusade in 1204, the Image of Edessa disappears from recorded history."[35]

According to Guscin, historians cannot definitively pinpoint the origin of the Image of Edessa. One well-known account of its origin, the *Narratio de imagine Edessena*, was presented at the time of the Image's arrival in Constantinople in 944. Guscin writes that "The *Narratio de imagine Edessena* dates the origins of the Image of Edessa to the time of Christ himself, shortly before his passion, as do the vast majority of other sources. It should be pointed out that the *Narratio de imagine Edessena* itself gives two possible versions for the origins of the Image, one the regular Abgar story (i.e. the king sends a messenger to paint a picture but Jesus miraculously imprints his facial features onto a cloth and sends it back to Abgar), while the second version stages the imprint story in the garden of Gethsemane; when Christ was sweating blood (Luke 22:43-44), he was handed a cloth to wipe his face on and the image of his face was miraculously transferred onto the cloth."[36] The same account acknowledges that the Image was hidden away for several centuries after a later king turned against the new religion. The true story of the Image's origin was thereby lost with only two elements surviving: the image dated back to the time of Christ and it was "not made by human hands."

The Image was famous enough for a tenth century Byzantine Emperor to launch a campaign to capture it from Edessa which was by then under Islamic rule. In return for financial compensation and the release of Moslem prisoners, the Islamic rulers surrendered the image to the Byzantines. It was received in Constantinople in a triumphant procession in 944. By the turn of the first millennium, Constantinople had become the largest and most glamourous city in the Western world. Even before that, it had become the relic-house of Christendom. Notably, in later years the burial linens of Jesus were listed in the official Constantinople catalog of relics.

What did the Image of Edessa look like? Most of what we know about this comes from descriptions given in Constantinople. But there are a few pre-944 reports as well.

> *Andrew of Crete* (660-740) – "[The] venerable image of our Lord Jesus Christ on a rag sent to Abgar, the toparch, which is an imprint of his bodily features."[37]

John Damascene (675/6-749) – "The Lord himself laid a cloth on his own, life-giving face, so as to imprint his image on the cloth, thus to send it to the longing Abgar."[38] John of Damascus speaks here of a himation, a full-length cloth.

Second Council of Nicaea – "In 787, during the Second Council of Nicaea, ... Leo, a reader of the Church of Constantinople, brought a personal testimony: "I've been to Edessa and I saw the holy image, not made by human hands, honored and venerated by the faithful."[39]

Alexis – Alexis was a holy man of Edessa who came from Rome and settled there possibly as far back as the fifth century. There are several lives of Alexis that speak of his connection to the Image, the earliest being an eleventh century French poem. In a paper published in the journal *Modern Philology*, Linda Cooper connects these descriptions of the Image with the Shroud of Turin. "The eleventh century Old French Life of Saint Alexis," she writes, "generally recognized as the first masterpiece of French literature, contains the passage:

> 'Then he [Alexis] went off to the city of Edessa
> Because of an image he had heard tell of,
> Which the angels made at God's commandment.'"

Cooper also quotes descriptions from two other lives of Alexis that indicate the Image has features found on the Shroud: the later Bollandist Vita Alexius which calls it "sine humano opere imago Domini nostri Jesu Christi in sindone," i.e. "an image of our Lord Jesus Christ made without human work on a sindone" [shroud]; also the Latin Cod. Monac. Aug. S.Ulr. 111 "Edisse [venit] in urbem, in qua sanguinea domini serva[ba]tur ymago non manibus facta", i.e. "[he came] to the city of Edessa, in which there was preserved a blood-stained image of the Lord not made by hands."[40]

So was the Image of Edessa the Shroud of Turin?

The best-known modern Shroud author Ian Wilson has used a multitude of data points to argue that the Image of Edessa is what we know today as the Shroud of Turin. Only a handful of academic historians have taken the time to seriously study this issue. Of these, some have rejected Wilson's thesis. Others have found his case compelling and have buttressed it with additional evidence. Wilson himself has responded to both actual and potential criticisms.

The main arguments against the identification of the two are well-known and have been articulated most eloquently by another Byzantine historian Averil Cameron (Turinese polemicist Andrea Nicolotti, also a critic, tends to substitute abuse for argument):

- Edessa was an image just of the face and not of a body
- Edessa is an image on a cloth the size of a handkerchief and not one of a full body on a shroud-size cloth
- Edessa shows Christ alive unlike the face on the Shroud
- Edessa was not associated with the Passion let alone the shroud of Jesus

The arguments in favor are also quite familiar.

- Most important of these is the point that the Image has been described as a *tetradiplon* (Acts of Thaddaeus, Menologion), a linen cloth folded in four. When the Shroud is folded in this fashion, with only the face visible, it looks much like ancient depictions of the Image of Edessa. Creases in the Shroud indicate it was thus folded.
- Although the Image has been described as being the size of a handkerchief, it was also spoken of as a himation or larger garment prior to its arrival in Constantinople. (e.g., John of Damascus)
- Likewise, in this same period, it has been referred to as a full-body image. (e.g., Pope Stephen)
- The image was spoken of as faint and moist much like the Shroud

- When the Image first came to Constantinople, the Emperor's sons specifically referred to their inability to see the eyes of Jesus. The eyes of Christ in icons based on the Shroud are shown as wide open and staring to emphasize the resurrection.
- As is the case with the Shroud, the Image was generally described as an imprint and not a painting
- With regard to the Passion, Kitzinger said, "By the end of the fourth century proskynesis [prostration] before the Sign of the Passion was considered a perfectly natural thing for a Christian."[41]

Averil Cameron had said, 'The form in the Image of Edessa story .. makes quite impossible the notion that the cloth in question could have been the extant shroud, even if we could believe the idea that it was folded (why?) for many centuries in such a way that only the face could be seen." [42]

In response, Wilson writes,

Now in the quotation from Professor Cameron's lecture you may have noticed her single, un-developed interjection "why" in respect of the concept of the Image of Edessa having been folded – as if there was no justification for such an idea. She confined to perfunctory mention in a footnote any discussion of what to me was the single most important breakthrough in the case for the Shroud being the Image of Edessa. I refer in this instance to the use of the Greek word [*tetradiplon*] to describe the cloth, first in the sixth century *Acta Thaddei*, then ... in the "Liturgical Tractate" of the 10th century. The point here is this: [*Tetradiplon*] is a most unusual word, occurring *only* in respect of the Image of Edessa in the entire corpus of Greek literature. But it is also perfectly understandable, being a compound of two very ordinary Greek words ... meaning "four" and ... meaning "doubled", hence "doubled in four".

And what happens when we try doubling the Shroud in four? We find precisely the face only arrangement that I have contended was the source of the Byzantine belief that the Image of Edessa was a face-only cloth imprint made while Jesus was alive. When we look at early artists' representations of the Image of Edessa we find the cloth represented in precisely the landscape-shaped format that arises from the Shroud being folded in this manner. We see the same cloth coloration. We see the same image coloration. In literary texts we find the Edessa cloth's image called ... an imprint, ... (acheiropoietos) "made without hands" and further described as "a moist secretion without colouring or painter's art". ...

The foldmarks are, however, a different problem. Even if the folding arrangement minimized stress, nonetheless one would expect pronounced crease lines after what would have been more than a thousand years in the same position – although with appropriate moistening, really ancient foldmarks can be smoothed out to a surprising degree. In fact, the Shroud's surface, when seen in an appropriate raking light, is literally criss-crossed with creases and foldmarks of all kinds, including from various folding arrangements historically known since the fourteenth century. These inspired Dr. John Jackson, in collaboration with photographer Vernon Miller, to make a special study of them as part of the STURP testing programme in Turin in 1978.[43]

In later years, in a comment on the Image of Edessa, Cameron admits that "in the Greek Acts of Thaddaeus of the seventh century or later, there is more detail: the cloth is called both tetradiplon and sindon, the latter word suggestive of a full-length cloth or shroud."[44]

In his anthology of all extant texts on the Image of Edessa, Guscin notes that "it should be stressed that there are no artistic representations of the Image of Edessa as a full-body image or with bloodstains, and the majority of texts make no reference to either characteristic; but at the same time it is undeniable that at some point in the history of the Image of Edessa, some writers were convinced, for whatever reason, that it was indeed a full-body image on a large cloth that had been folded over (possibly in such a way that only the face was visible), and that it did contain bloodstains."[45]

It must be remembered that accounts of the Image's history say that it was re-discovered after being hidden for centuries. Those who rediscovered it could not be sure of its precise origin beyond its association with Jesus – hence the various Abgar stories. All we really know is that Edessa had an image of Jesus "not made by human hands" and that it had been "called both tetradiplon and sindon, the latter word suggestive of a full-length cloth or shroud" (to quote Cameron). Edessa, it would seem, is a bright object that bespeaks the presence of a hidden galaxy. But what we see through a glass darkly in Edessa, we see face to face in Constantinople. So let us journey to the heart of Byzantium.

The Constantinople Constellation

As we have seen, the Image of Edessa was brought to Constantinople. But whatever might be said about the Edessa-Turin connection, it is a matter of historical record that Constantinople claimed to possess the shroud of Christ. Was the shroud of Constantinople the Shroud of Turin?

Here a statement of the historian Ernst Kitzinger is especially relevant. He is reported to have identified the Shroud of Turin as the Shroud of Constantinople. Kitzinger was asked "'Can you show me some works of artists who have painted blood marks like the ones that you see on the Shroud of Turin?' His reply: 'The Shroud of Turin is unique in art. It doesn't fall into any artistic category. For us, a very small group of experts around the world, we believe that the Shroud of Turin is really the Shroud of Constantinople. You know that the crusaders took many treasures back to Europe during the 13th century and we believe that the Shroud was one of them. As for the bloodmarks done by artists, there are no paintings that have blood marks like those of the Shroud. You are free to look as you please, but you won't find any.'"[46]

Likewise, Hans Belting, the eminent historian of Byzantine art cited earlier, said that the Shroud of Constantinople (referred to as the sindone by Robert of Clari) was "probably identical with the 'Shroud of Turin.'"[47]

The focus here is not on the Image of Edessa (we will separately consider its connection to Constantinople) but on the undeniable presence of the Shroud in Constantinople from the tenth century. Undeniable because Constantinople claimed to have the Shroud; witnesses from entirely different kinds of backgrounds testified to the presence of an artifact that sounded very much like the Shroud; and, of course, paintings that looked like the Shroud full-body image began to trend.

This ground is best covered by professional historians. One of the most prolific historians in this domain is Dan Scavone, author of numerous works on the historical background of the Shrould of Turin. Particularly helpful is his systematic citation of sixteen documents linked to Constantinople that refer to a Shrould-like artifact, excerpts from which are cited in Appendix 2. Especially relevant excerpts are given below.

Introducing this study, Scavone writes:

> Numerous documents describe in important detail the presence in Constantinople of an icon of Jesus's face on a cloth which in the year 944 had come from the city of Edessa, modern Urfa in southern Turkey. This icon, known also as the Mandylion, was said to be miraculously imprinted, a likeness not made by human hands, or *acheiropoietos*. ... I have selected sixteen of these documents for close scrutiny. The documents span the period 944 to 1247. Four of the earliest documents, datable from 944 to 960, refer to the Mandylion alone. Six others, those dating from 1150, 1200, 1201, 1203, 1207, and 1247 also assert the presence in Constantinople of Christ's burial wrapping, or portions thereof, along with the Mandylion. Six different documents from 958, c. 1095, 1157, 1171, 1205, and 1207, attest the burial wrappings but not the face cloth (Mandylion).

The emphasis upon a singular imaged cloth icon considered to be the actual burial wrapping in this study of acheiropoietos Jesus images is appropriate chiefly because one most important document of 1203, the memoir of Robert of Clari, a knight of Picardy, reported seeing "the burial cloth (sydoines) with the figure of the Lord on it."

Scavone addresses the problem of "the time of the arrival in the capital of the reputed burial shroud icon of Christ. Whereas the Mandylion was received in Constantinople with a great celebration (Documents I and III), not a single source records the arrival there of any larger Jesus-icon. It is, however, included in a number of documents, as already noted, and at least once explicitly with a Christ-image on it. The implication of this is that the Mandylion and the burial wrap icons may prove to be one and the same object."

The identity between the two becomes apparent in Constantinople.

"A development of the 10th century, one clearly associated with the Mandylion's arrival in the capital and its accessibility to new and more sophisticated eyes, was the revelation in the two eyewitness sources produced immediately upon its arrival that the icon also had blood on its face and, surprisingly, that it had a bloodstain where Jesus had been stabbed in the side while on the cross. In light of these data and recalling the term *sindon* of the *Acts of Thaddaeus*, we may rephrase this third question: Could the Edessa Mandylion always have been a folded burial shroud icon now assumed in these Constantinopolitan sources to be the real blood-stained burial wrapping of Jesus, whose separate arrival in the capital is nowhere mentioned? This initial awareness of larger size and of blood on the Mandylion is the thrust of my first two documents." [see Appendix].

Scavone concludes:

To sum up the points made in this paper: a linen cloth or cloths described as the burial wrappings of Jesus are attested in many Constantinople documents from 944 to 1203, twice with his image if one counts [Nicholas] Mesarites (Doc. XI), and several times described as bloodied. No record exists of the arrival of Jesus' burial cloth in the capital, and no celebration such as accompanied the Edessa cloth in 944. Yet it was there. Judging from copious documents and artistic representations made in Constantinople and elsewhere from 944 to 1150, the Edessa towel always with the image of Jesus' face may be identical with Jesus' Shroud in folded form, enclosed in a case with face exposed. Before that, from at latest 544 to 944, this cloth was certainly in Edessa. If the Edessa cloth and Jesus' purported shroud are indeed one and the same object, that assumed burial cloth may have a pedigree back at least to 544, and if the Abgar legend has any historical worth, to the 4th c. and even, accepting the descriptive evidence, to the very time of Christ. If the pieces of this elaborate puzzle truly fit as they seem to, the blood-stained burial cloth with faint unpainted image would have a documented history back to palaeochristianity and may in fact be the actual tomb wrapping of Jesus.[48]

Is the Image of Edessa the Shroud of Constantinople that later became the Shroud of Turin? It is clear that the story of the Image of Edessa seemed to be morphing in the direction of the Shroud. There was an underlying logic to this movement as Edessa specialist Mark Guscin explains:

> On 15th August 944, the Image of Edessa, the … image (not made by human hands), came to the imperial capital Constantinople from Edessa (today's Sanli-Urfa in Turkey). … A sermon pronounced by Gregory Referendarius, Archdeacon of Hagia Sophia in Constantinople on the occasion of the Image's arrival in the city survives in one known manuscript in the Vatican Archives, recently rediscovered by Italian classics scholar Gino Zaninotto. …

The main objection to the theory [that the Image of Edessa is the Shroud of Turin], still made by many today, is that the Image of Edessa is generally recorded as a facial image of Christ formed in life, either when he met the messenger sent by King Abgar and wiped his face on a cloth, miraculously leaving an imprint on it, or when he wiped his face with a cloth in the garden of Gethsemane, while sweating blood. This second theory of the legend of the image, found in the Official History attributed to Constantine Porphyrogenitus, written on or shortly after the arrival of the Image in the imperial capital in August 944, is proof that blood had been seen on the Image as the author was trying to find an explanation for this.

What seems to have happened is that each time something new was discovered about the Image of Edessa, it was incorporated into the legend with no attempt to iron out any resulting contradictions. Yes, the image on the cloth was always said to be a facial image formed in life, but in some texts alongside this statement we also read that the image was of the full body of Christ. Gregory himself, in the sermon translated below, attributes the image to the sweat from the agony in Gethsemane, while at the same time declaring that the image has been embellished by blood from the side wound (quite obviously after the crucifixion and the death of Jesus). Of course this kind of internal contradiction makes our twenty-first century minds jolt and think of explanations, such as interpolations, texts added on or even invented. And yet our minds do not work like those of devout Orthodox believers. ... According to some hymns contained in the Menaion, when Abgar contemplated the cloth, he saw bloodstains on it, and yet the hymns make no attempt to explain this referring to Gethsemane. New information is included without eliminating or correcting what was previously thought to be true. ...

Arguing that the Image of Edessa cannot be identified with the Shroud of Turin just because there are plenty of texts that describe the Image as a facial image formed in life is rather like arguing that the planet Pluto does not exist, and bringing out a pre-1930 encyclopedia to prove the point. The only difference is that post-1930 books rectified the point in question, whereas the many Abgar texts just included the new information alongside the old (as if an encyclopedia were to say there were eight planets and then list nine). No matter how unsatisfactory this might be to our way of thinking, this is how the texts present the story. The Gregory Sermon is precisely one of these texts. The author quite obviously knew that the Image of Edessa had both bloodstains and a side wound. He does abandon the idea of Jesus pressing a cloth to his face in reply to Abgar's letter and messenger, because he knows there is blood on the cloth – the slightly later Official History of the Image of Edessa offers both possibilities, again aware that there is blood on the cloth. He therefore concludes that the image must have been impressed onto the cloth when Jesus' sweat ran down his face like drops of blood in the garden of Gethsemane.

However, this would not explain the blood from the side wound. Gregory therefore assumes that "afterwards" (giving a temporal sense to the Greek...) the Image was embellished with blood from the side wound. At no point does he even consider the possibility that the Image of Edessa was in fact the burial shroud of Jesus Christ, once again, something that from our own point of view, seems quite obvious. He must, however, have assumed that the cloth used in Gethsemane was again present at some moment shortly after the crucifixion. Gregory's lack of inquisitive spirit about the origin of the image and the side wound is most frustrating, yet no matter what he did or did not think about the origin of the blood from the side, one thing is clear – according to this sermon, the Image of Edessa had a bloodstain from the wound inflicted on Christ's side, and therefore contained a full body image. No amount of contrived pseudo-translations or explanations can get away from this simple fact. This is therefore a text of great relevance and importance for the history of the Image of Edessa and its possible identification with the Shroud of Turin. Apart from the sixth or seventh century reference to an imaged and bloodstained burial shroud in the Old Spanish liturgy (also called the Mozarabic Liturgy), there are numerous texts that describe the Image of Edessa as much more than a simple facial depiction of Christ. This text, translated below for the first time into English, describes blood from Christ's side, a very strong piece of evidence for identifying the Image with the Shroud.[49]

The three most famous and unmistakable references to the Shroud in Constantinople came from Gregory Referendarius (944), whose sermon has already been referenced, Nicholas Mesarites (1201) and Robert de Clari (1203). Gregory and Mesarites were prominent Byzantines, Gregory being a high-ranking cleric of the patriarchal basilica of Constantinople and Mesarites the custodian of the holy treasures of Byzantium. de Clari, on the other hand, was a French knight who came to Constantinople in 1203-4 as a part of the Fourth Crusade.

Mesarites specifically includes the burial cloth of Jesus in his list of the relics in Constantinople: "The burial shrouds of Jesus; these are of linen, an inexpensive fabric, which still smell of myrrh; they are not subject to alteration because they covered the Dead One, clearly nude, and were full of myrrh after the passion. ... In this place he rises again and the sudarium and the holy sindons can prove it."[50]

Describing the marvels he witnessed in Constantinople, De Clari writes: "There was another church which was called My Lady Saint Mary of Blachernae, where there was the SYDOINES [sheet] in which, Our Lord had been wrapped, which every Friday, raised itself upright, so that one could see the form [figure] of our Lord on it, and no one, either Greek or French, ever knew what became of this SYDOINES when the city was taken."[51]

In his commentary on these and other utterances, de Wesselow observes that "Sindonologists have long regarded this [de Clari's description] as a likely reference to the Shroud. The word Robert uses to denote the cloth, *sydoines*, is simply an Old French spelling of the Greek word *sindon*, meaning a "linen sheet," the word used in the Synoptic Gospels for the cloth in which Jesus' body was wrapped. This *sindon* is identified as the actual burial cloth of Christ, and, crucially, it is said to have manifested the *figure*, i.e. the bodily form, of Christ. The description matches the Shroud. And Robert says that the *sindon* disappeared after the sack of the city, presumably looted by a Crusader, which may help to explain how the Shroud ended in France. ... Taking the historical evidence on its own, it is perfectly reasonable to connect the cloth seen by Robert de Clari with the Shroud. If Robert's *sindon* was the Shroud, it would predate 1260 – the earliest date indicated by the carbon dating – by over half a century."[52]

About Mesarites' statement, he notes,

> There can be no doubt that Nicholas is referring to the same relic as that seen by Robet de Clari. He calls it the *sindones*, just as Robert refers to it as the *sydoines* – the same word in Old French. Both witnesses identify it as the linen in which Jesus was wrapped, and there cannot have been more than one cloth claiming this distinction at the same time in the city.
>
> If the cloth relic described by Nicholas Mesarites was one and the same as that seen by Robert de Clari, then it would follow that it also manifested "the figure of Our Lord." Although Nicholas omits to mention this crucial point, his words do evoke the Shroud in several specific and surprising ways.
>
> First, Christ's body is referred to as naked. This is significant, as in this period the dead Christ was almost invariably conceived as wearing a loincloth (cf. Figures 23 and 24). The novel idea that Christ was naked when wrapped in his winding sheet could have been inferred from the Shroud. Secondly, the adjective *aperilepton*, meaning literally "unoutlined," is a word that is obviously applicable to the blurry, un-outlined Shroud-image. What better way of describing the figure seen in the

Shroud? Thirdly, the odd remark about the cloth defying destruction might hint that it was conspicuously damaged. This would fit in with evidence to be adduced below that the so-called "poker-holes" were burned into the Shroud before the thirteenth century.

On its own, Nicholas's account of the Sindon would mean little, but in combination with Robert de Clari's account it is highly significant. Together, these two reports indicate that, at the beginning of the thirteenth century, a (possibly damaged) linen sheet was kept in Constantinople that bore a blurry image of the naked, crucified Jesus. This sounds like a description not just of a burial cloth relic similar to the Shroud, but of the Shroud itself.[53]

Why didn't Mesarites speak directly of the image. De Wesselow explains:

But why would the Overseer have failed to mention the all-important image? The answer may have to do, at least in part, with the problematic discovery of the relic. Originally, as we shall see, the cloth was probably framed and interpreted as something else entirely, a miraculous portrait of Jesus, an image whose cult was too important to be discredited or compromised. There was also a more general reason: religious mystique. That the Pharos Chapel housed the burial cloth of Christ seems to have been common knowledge, but awareness of the "miraculous" image on the cloth appears to have been reserved to a privileged few. Byzantine society was extremely hierarchical, and the awe-inspiring image was probably deemed too sacred a sign to be shared with hoi polloi. Before the Crusaders arrived on the scene in 1203 there would have been no question of advertising the existence of the image, let alone displaying it in public. Had Nicholas Mesarites mentioned it in 1201, he would have betrayed a royal and aristocratic secret.[54]

So how old is the burial cloth?

> How far back can we trace the Byzantine relic of the Sindon? The references are few and far between. The chronicler William of Tyre records the Sindon among various relics shown to King Amaury of Jerusalem and his entourage in 1171. Going further back, a letter of 1092 purporting to be from the Byzantine emperor to various Western princes tells us that "the linen cloths [*linteamina*] found in the sepulchre after his Resurrection", were then in Constantinople. Although the source is somewhat problematic, this is presumably a reference to the same relic. The Sindon is first mentioned more than a century earlier, in a letter of encouragement sent by Emperor Constantine VII Porphyrogenitus to his troops in 958. The emperor says that he is sending them some holy water consecrated by contact with various relics of the Passion in the Pharos Chapel, including the *theophoron sindonos* – the "God-worn linen sheet." Whatever the precise meaning of the word *theophoron*, this is a clear sign that the Sindon seen by Robert de Clari was in the imperial relic collection by the mid-tenth century – a full 300 years before the earliest date indicated by the carbon dating of the Shroud. [55]

But was the Shroud of Constantinople the Shroud of Turin?

> The million-dollar question remains: can we be sure that the Sindon of Constantinople was one and the same as the Shroud of Turin? The Byzantine cloth certainly seems to have been a close match for the present-day Shroud, but is there any evidence that proves they were identical? Indeed, there is. But it is not a written description: it is a rather crude drawing in a medieval manuscript.

> One of the greatest treasures in the National Library of Budapest is the Pray Codex, a manuscript which contains, among other things, the earliest Hungarian annals and the earliest work of Hungarian literature. The main part of the codex, including the

part that interests us here, was produced around 1192-5 in one of the country's Benedictine monasteries. ...

On folio 28r of the Pray Codex, in the midst of a liturgical text relating to the celebration of Holy Week, are a couple of drawings that together document the existence of the Shroud in the late twelfth century. ...

The upper scene is a rare depiction of the Anointing of Christ ... [which] corresponds to the Shroud ... in five telling respects: it represents Jesus naked, his wrists crossed over his groin, his hands lacking thumbs, a prominent red stain above his right eye, about to be enfolded in a long sheet drawn up and over his head. Does this not look like an attempt to imagine the burial of Christ on the basis of the Shroud? What are the odds in favor of all these rare correspondences with the Shroud occurring in the same image just by chance? ...

[In all there are] eight telling correspondences between the Shroud and the drawings on a single page of the Pray Codex. The first five, found in the scene of the Anointing, are sufficient on their own to indicate that the artist of the Pray Codex knew the Shroud. Conclusive proof is provided by the three correspondences in the lower scene: the stepped-pyramid pattern in the upper rectangle, evoking the distinctive herringbone weave of the Shroud; the folding of the object in two halves; and the small circle formations, which match the pattern of the poker-holes. It is inconceivable that all these detailed links with the Shroud, several of which are found nowhere else, could have occurred on a single manuscript page by chance. The only reasonable conclusion is that the artist of the Pray Codex was aware of the Shroud.

The Shroud existed and was already damaged, then, by 1192-5, when the illustrations in the Pray Codex were drawn. Given the close links at the time between Hungary and Byzantium, it can hardly be doubted that the artist saw the relic in Constantinople. The Shroud was the Byzantine Sindon. ...

The Shroud of Turin, then was once the Sindon of Constantinople. Seen in public by Robert de Clari and his fellow Crusaders in 1203-4, it was kept before then in a state of religious purdah, witnessed only by members of the Byzantine court and esteemed visitors, who, once initiated, could be trusted to keep the secret of its astonishing image. Historical records show that the Sindon was kept in the Pharos Chapel as part of the imperial relic collection, being first documented there in 958, 400 years before it was put on show in the small French village of Lirey.[56]

De Wesselow also addresses the variations in the Edessa origin-story:

One of the chief characteristics of the Mandylion was that it 'bred' copies of itself, which is to say that numerous copies were made, some of which were seen as having come into being miraculously. And this may help explain why, by the beginning of the 13th century, the 'Mandylion' and the 'shroud' were understood to be two separate relics in the Pharos Chapel [in Constantinople]. Discovering that the famous face-cloth sent to King Abgar was actually a burial cloth would have presented the relic's custodians with a problem, and the easiest way to solve the problem was to transfer the story about King Abgar to a copy of the facial image. This may have been deliberate, or it may have happened accidentally.[57]

Appendix 2 documents historian Daniel Scavone's chronicle of contemporary reports relating to the Shroud of Constantinople.

"I don't believe the Shroud is a medieval work of art because I don't believe in miracles"

Assuming that the Shroud was in Constantinople, what happened after the city was sacked by the Crusaders? Much ink has been spilled over theories of the route taken by the Shroud from Constantinople to Lirey over a period of a century and a half. There are various plausible accounts involving the Knights Templar, the King of France, and others. But there is not much to go on other than speculation.

We have three hard facts:

- Speaking of the looting of Constantinople, Theodore Angelus said: "the French did the same with the relics of the saints and the most sacred of all, the linen in which our Lord Jesus Christ was wrapped after his death and before the resurrection."[58]
- Concerning the same matter, Nicholas Mesarites, wrote, "When the city was captured by the French knights, entering as thieves, even in the treasury of the Great Palace where the holy objects were placed, they found among other things the .. spargana/*fascia*." (the burial linens of Jesus).[59]
- The Shroud was brought to Lirey by a French knight with a name similar to another knight who had been in the Crusades.

This is what we know and taken together it seems like fairly persuasive circumstantial evidence connecting the Shroud of Lirey to the Shroud of Constantinople.

De Wesselow notes that Theodore Angelus, in his letter mentions that the stolen Shroud was initially taken to Athens. This is important because "The connection with Athens may help explain the eventual re-emergence of the cloth, a century and a half later, in rural France. The Lord of Athens in 1205 was a crusader called Otho de La Roche, who was a direct ancestor of Jeanne de Vergy, the woman who, along with her husband, Geoffrey I de Charny, staged the first exhibition of the Shroud in France in 1355/6. This suggests that the Shroud may have been a secret family heirloom passed down from the man who looted it in 1204."[60]

An obvious question is why Geoffrey I de Charny did not announce the source of the artifact. The obvious answer is that its latest owners were holding stolen property and had no desire to alert its original owner.

De Wesselow expands further:

> Suspicion has also been aroused by the failure of the cloth's owners, the de Charnys, to explain how they came by their extraordinary treasure. They did no more than offer conflicting hints, Geoffrey II de Charny saying that his father had been given it, Geoffrey II's daughter, Margaret, saying that Geoffrey I had won it as a spoil of war. This shiftiness has been taken as a sign that the de Charnys knew the relic was a fake, but it might just as well indicate that their ownership of the cloth was somehow illegitimate. Having identified the Shroud as the Sindon of Constantinople and remembering the Fourth Crusade, we can now see that this was the case. The de Charnys could not divulge the provenance of the Shroud or openly declare it to be the true Shroud of Christ, because it was not rightfully theirs – or any other Westerner's. They would have risked having it confiscated. It was preferable to pay lip-service to the idea that it was a copy, maintain possession of the cloth and look for a future opportunity to promote its cause.
>
> The Shroud's problematic provenance also explains why Pope Clement VII permitted displays of the cloth to continue (to the dismay of Pierre d'Arcis), but only if it was publicly proclaimed to be "a figure or representation of the Shroud of Our Lord," not the real thing. As a relative of the de Charnys, Clement almost certainly knew the cloth's provenance, but he could not allow it to be recognized as the true Shroud of Christ, for fear of causing a diplomatic incident. The Shroud was a cultural treasure that meant as much to the Greek emperors of Byzantium as the Elgin Marbles do to Greeks today, and it had been stolen from them in a looting campaign as brazen as any perpetrated by Napoleon or the Nazis. Nearly two centuries

after the Fourth Crusade, the Sack of Constantinople was still an extremely sore point in Byzantium, and, if John V Palaiologos, the Byzantine emperor at the time, had heard that the priceless Sindon was being displayed in France, he would have made moves to recover it. Pope Clement would have been put on the spot, and the anticipated reunion of the Roman and Byzantine Churches, which was being actively discussed at the time, would have been put in jeopardy. Clement, then, had good reason not to acknowledge the real identity of the Shroud and to enjoin perpetual silence on Bishop d'Arcis as well. [61]

Speaking of Bishop d'Arcis, there is also the problem of his charge that the Shroud was a painting. Here again de Wesselow's analysis is invaluable:

The case against the Shroud rests largely on a dubious reading of the documents concerning its emergence in the mid fourteenth century. To begin with, skeptics have leaped on the claim of Pierre d'Arcis, bishop of Troyes, that "thirty-four years or thereabouts" before he wrote, i.e. in about 1355, one of his predecessors, Henry of Poitiers, had investigated the Shroud and found it to be "a work of human skill." In the late nineteenth century, when scholars first became aware of this claim, they were understandably inclined to take it seriously, since little or nothing was then known about the Shroud and its image. In the early twenty-first century, however, there is no longer any excuse for believing the Shroud to be a fourteenth-century work of art. Enough is now known about it (and about fourteenth-century art) to render the idea risible. We might as well dismiss the fossils of archaeopteryx found in Germany from the 1860 on as fakes on the basis that they were denounced as such by a few contemporary scientists. Rather than swallow Bishop d'Arcis's claim, we should recognize it as hearsay evidence that tells us something about fourteenth-century events at Lirey, but nothing about the origin of the Shroud.

There is no substance to Chevalier's claim that the fourteenth-century documents concerning the Shroud at Lirey prove that it was a recently executed painting – or any other sort of artwork. This was a conclusion reached in ignorance of the Shroud's exceptional qualities as an image and its early, eastern history, proved by the Pray Codex. The poker-hole patterns represented in the Pray Codex drawing, first noticed in 1998, are also the final nail in the coffin of the carbon-dating result. The cloth now in Turin must be at least three centuries older than the earliest date indicated by the radiocarbon age of the sample tested – a sizeable error. The 95 percent confidence level the laboratories cited is meaningless, except, perhaps, as a measure of scientific hubris. Physics is not the only way to date the Shroud; historical and art-historical records have their part to play, as do the various indications gleaned from medical, chemical and archaeological investigations. This broad spectrum of research indicates that the Shroud dates not from the Middle Ages, but from antiquity.[62]

Scavone has also addressed the D'Arcis memorandum:

[D'Arcis: "About thirty-four years ago [about 1355] an inquest was held into the Shroud. Expert theologians [NB: not artists] concluded the Shroud was false because no image is mentioned in the Gospels. Also, the artist came forward."]

The Bishop's words, "about thirty-four years ago," suggest that he had no dated document before him and had no first-hand knowledge about the artist. Nor do any of the Pope's responses to D'Arcis refer to an artist. The bishop says later that he has been accused of desiring the Shroud for himself and has become a "laughing stock." In fact, the collapse of the nave of his cathedral of Troyes in that very year, 1389, had resulted in the loss of its most precious relics, magnets for pilgrims and their contributions. This creates a presumption of self-serving in his memorandum to the Pope.

One might add that if an artist had produced the Turin Shroud in the fourteenth century, he would have been an original, creative genius of the first magnitude for his realistic rendering of anatomy and bloodflows, beyond anything known in Gothic art. He would have created the first nude Christ. His idea of a double image on a cloth would be unique in the history of Christian art. The Shroud does not, in fact, fit in the context of any artistic style or genre. For its realism, if it were art, it would claim a place on page one of every book on the art of the Renaissance.

In fact, a "shroud" was painted about that time. The Besanton shroud was painted in the 14th c. and it is to this painting on cloth that the D'Arcis memo may refer. The artist may indeed have come forth. This painting (which exists) was likely made to substitute for the real Shroud which "disappeared" from Besanton after the fire of 1349 which destroyed its cathedral. The perpetrator of the disappearance, it may be argued, was Jeanne de Vergy, daughter of one of the most prominent families of that city, who, about 1353 married Geoffroy de Charny, first known owner of the Shroud. One may argue that she brought that object with her as a sort of dowry to her marriage.[63]

There are other kinds of doubts about the D'Arcis Memorandum as well:

"In the 1990s, Parisian researchers determined that the so-called 'D'Arcis memo' was no memo at all, but merely a clerk's draft in poor Latin, never dated nor signed nor sent to the Vatican, and with no official copy in either Troyes or the Vatican archives."[64]

As a professional art historian, de Wesselow finds it impossible to locate the Shroud in the stylistic setting of medieval art. *It would be nothing short of a secular miracle for this to happen.* But, as he told us in conversation, as an agnostic he does not believe in miracles. To quote him, "Personally I do not believe that the Shroud is a medieval work of art because I do not believe in miracles."

Similar strictures apply to other attempts to dismiss the Shroud as a painting.

Charles Freeman has floated the idea that the Shroud originated as a prop in a medieval liturgical performance of the empty tomb narratives ("Quem Quareitis"). According to him, the Shroud, as we have it, was much brighter and clearer in its first days but the painting (along with such accoutrements as a loincloth) has faded away over time with the constant rolling and unrolling. He also says that there was a new fascination with bloody scenes of the Passion in the Middle Ages and the image of the Shroud fits in with this trend.

Freeman's thesis has been challenged on various fronts. In the first place, as noted in a publication of the British Society for the Turin Shroud, the whole Quem Quareitis argument was addressed over two decades ago by Ian Wilson. Secondly, the pigments required to substantiate the painting hypothesis are not found on the Shroud. As the BSTS put it, you can argue they disappeared over time but it seems a bit much for the disappearance to be so uniform. There is no rabbit skin collagen either because the only protein found by STURP is from the bloodstains. Yes, there is calcium carbonate on the Shroud as one would expect from a cloth kept in a tomb carved out of limestone. Thirdly, if it was a painting and had been folded as often as the Shroud is known to have been, the image would have been damaged.

De Wesselow has his own comment about Freeman's theory:

> Freeman claims that "the Shroud was produced in medieval times 'by a method that we still have not worked out.' In other words, he has no idea what the Shroud is, but he presumes it must be medieval, based on the carbon-dating. Can he point to any other medieval work of art that a century of modern research has failed to explain in any way whatsoever? Can he explain away the plentiful evidence that the Shroud is the burial cloth of a man crucified by the Romans as a would-be king and buried in the manner of a Jew? The conviction that the Shroud is a medieval fake is like a religious belief that can't

be explained rationally and is held in the face of a great deal of evidence to the contrary. It is an article of faith that many hate to have challenged."[65]

Perhaps the most powerful critique of the Freeman theory came from fellow-skeptic Colin Berry:

The Shroud of Turin is certainly not a painting. The mainly US-based STURP (Shroud of Turin Research Project) task force showed as much in 1981, searching for but failing to find known artists' pigments, bar a few flecks of iron oxide (artists' red ochre?). ... Getting the right words to describe the Shroud image into the media and public domain has acquired a new urgency of late, given the recent claims that attempt to undo decades of research. I refer to historian Charles Freeman's claim that the TS is merely an age-degraded painting. I've said quite a lot on that score already elsewhere, as indeed have others, and have little more to add, except to say that Mr. Freeman needs to get up to speed with Shroud science, and disabuse himself of the idea that it's all about art history. The TS is arguably NOT about art. It's an artefact, intended for purposes other than mere artistic expression. ...

It's no longer sufficient in this blogger's view to continue describing the TS as a "faint image." That is too non-specific and makes it too easy for CF to peddle his antediluvian views... "Faint image" or even faint NEGATIVE image simply does not do the business (CF having closed his eyes completely to the implications of the tone-reversal implied by the descriptor "negative"). No, we need new updated terminology that makes it clear that the TS is not just any old "faint image", but one with very special, indeed unique properties that sets it apart from other pictorial representations of the human form.

So what is that terminology to be? One has to be neither pro-nor anti-authenticity to regard the TS image as an IMPRINT.

Freeman responded by adding qualifications to his theory but Berry would have none of it. He observed that Freeman now had to make "late-in-the-day qualifying assumptions" that "involves some wild speculation about pigment leaving shadows." Berry concludes: "This is not science. It is not even vaguely scientific. It's an attempt to dress up a dud hypothesis with ever increasing layers of fantasy."[66]

Another approach has been to dismiss the Shroud as one of many copies of the shroud of Jesus. This line of attack comes from Antonio Lombatti, another Italian gladiator whose chief weapons of combat are polemical wisecracks and fuzzy speculation. He first swears allegiance to the C14 study, Walter McCrone and Bishop D'Arcis. He then announces that there were some forty shrouds in circulation almost all of which were destroyed in the French Revolution. The shroud we have was simply the winner in a contest of the survival of the fittest. But all this is not only old news but about as relevant as the fact that there are many copies of Da Vinci's Last Supper. It is well known that there were numerous replicas of the Shroud: remember these copies were often touched to the original Shroud to be "sanctified." Remember also the Pray Codex. In 1952, Doménico Leone did a systematic study of the copies of the Shroud made before the modern era. More recently, Daniel Duque Torres carried out an updated version of this study in the *British Society of the Turin Shroud Newsletter* and notes that "There are copies the same size as the original, some very small ones (just 10 cm long), others with the spear and nail wounds in different positions, some with a crown of thorns and others without it, some from the same workshop and others absolutely anonymous. Some have texts written on (in Latin, French, Spanish and Italian), all of which leads us to the conclusion that the tradition grew over the years and at some times even got out of control."[67] Over eighty are currently in Europe and some forty were lost.

What is devastating as it applies to Lombatti is that all these shrouds were replicas and copies and understood to be such. More important, none have manifested the attributes unique to the Shroud of Turin. Oracular proclamations and grand hypotheses must in the end be judged by the supporting evidence. Not only does all the evidence point to the antiquity of the Shroud but there is no other candidate for the claim to be the burial cloth of Jesus. Lombatti is right, however, on one point. The Shroud is a survivor par excellence. It has survived fires and wars. And it will continue to survive the slings and arrows of polemicists and partisans.

Starstruck

De Wesselow notes that "It is clear, in any case, that the Shroud's history does go back, one way or another, to the first century – and to the tomb of Jesus. Enough was known about the Shroud in 1902, when Delage stood up and gave his pioneering paper at the Academie des Sciences in Paris, to make this a valid conclusion. A hundred and ten years later a huge amount more is known about the Shroud's physical make-up, image and history, and the conclusion is now far more secure: the only coherent way to understand it is as the burial cloth of Jesus."[68]

Returning to the historical question with which we started, we realize that, if nothing else, our inquiry into the history of the Shroud has clarified the issues involved. The data available are of different kinds. The conclusions possible are governed by the nature of the available data. We can neither exaggerate nor belittle what is possible in this regard.

The artifact we have in hand is extraordinary enough to suggest that it had an extraordinary beginning. If it really reaches back to the very origin of Christianity then its history would mirror the history of Christianity itself. Much of this history, we forget, was hidden. Persecution, conflict, the Dark Ages, plunder and pillage. But like a hidden star whose presence we detect from its gravitational pull, the Shroud has "shown" its presence from the very beginning: from early preachings on the shroud of Christ to the Big Bang of Christian art to the prominence of the Image of Edessa to the hallowed shroud of Byzantium to the emergence of the Shroud of Turin by way of Lirey.

Many details are disputable but the central elements are undeniable:

- From the very beginning there were accounts of an image of Jesus that was a direct imprint of his face and/or body
- The explosion of images of Jesus similar to the Man of the Shroud coincide with the recovery of this image of Jesus in the fifth century
- The image "not made by human hands" is eventually transported to the glory that was Byzantium
- Accounts of the relic treasury of Constantinople unmistakably indicate the presence of an image that can be identified with the Man of the Shroud
- French knights stole the Shroud of Constantinople and a French knight had it in his possession in France.

Now these elements taken in conjunction with everything else we know about the Shroud seem to furnish conclusive evidence that the Shroud of Turin is indeed the burial cloth of Jesus. But judgments are made by intellects. Reasonable people can disagree on where to go from these data points. Professional historians cannot be faulted for refusing to go beyond the strict canons of what is knowable from strictly historical evidence. But there is no compelling reason why other historians as well as non-historians cannot synthesize these very same data-points with what is known from other disciplines to generate Big Data insights. The historical data provide enough traction to taxi down the runway. The data from other disciplines provide the lift and thrust for the plane to take off. Those who wish to stay on the runway or nurse a fear of flying (!) cannot be forced to join the flight. But neither can they cancel the flights of those who are ready and willing to explore the friendly skies.

Coming back to the earlier metaphor, the star that is the Shroud beckons us forward. Your starlight from bygone eras, sometimes dim, sometimes dazzling, makes sense of your presence here and now. Equally, your starlight past and present makes us wonder how you are!

2.8 Authenticity and Antiquity

Dating

One final question: can we scientifically determine the date of the Shroud? The radiocarbon dating investigation could not help us given its demonstrable inability to decontaminate the samples being dated. In fact, any method for effectively decontaminating the Shroud would be so severe that no testing sample from it would survive. Thus, any new radiocarbon tests of Shroud samples would be exercises in futility. And there is no sense in seeking yet another C14 test because, as has been well said, insanity is doing the same thing over and over again and expecting different results.

But are there other alternative dating methods? Two other initiatives focused on the content of the Shroud fibers may be relevant but neither has won widespread acceptance.

The first was mentioned earlier, Raymond Rogers' measurement of the vanillin in the Shroud. His study was published in the peer-reviewed journal *Thermochimica Acta*. A BBC report on his work notes that "microchemical tests - which use tiny quantities of materials - provided a way to date the shroud. These tests revealed the presence of a chemical called vanillin in the radiocarbon sample ... but not the rest of the shroud. Vanillin is produced by the thermal decomposition of lignin, a chemical compound found in plant material such as flax. Levels of vanillin in material such as linen fall over time. 'The fact that vanillin cannot be detected in the lignin on shroud fibres, Dead Sea scrolls linen and other very old linens indicates that the shroud is quite old,' Mr [Raymond] Rogers writes. 'A determination of the kinetics of vanillin loss suggests the shroud is between 1,300 and 3,000 years old.'"[1] Critics have observed that vanillin content is not necessarily a reliable indicator of age so this initiative has had limited traction.

More interesting is Italian scientist Giulio Fanti's attempt to determine the age of the Shroud fibers using infrared light (Fourier Transform Infrared, or FTIR) and Raman spectroscopy. The age of the fibers was determined on the basis of the amount of cellulose in them. Fanti's study was published in the July 2013 issue of the peer-reviewed journal *Vibrational Spectroscopy*.

"The possibility to define a two-way relationship between age and a spectral property of ancient flax textiles has been investigated in the present paper employing both FT-IR and Raman analyses on selected samples dated from about 3250 B.C. to 2000 A.D. " says the report on the study, "For the first time, the possibility to define a correlation among spectral properties and age of flax samples, by using calibration curves, has been proved." [2]

The study and its conclusions were summarized in a popular Italian news site:

"The research includes three new tests, two chemical ones and one mechanical one. The first two were carried out … using infra-red light, and the other using Raman spectroscopy. The third was a multi-parametric mechanical test based on five different mechanical parameters linked to the voltage of the wire. The machine used to examine the Shroud's fibres and test traction, allowed researchers to examine tiny fibres alongside about twenty samples of cloth dated between 3000 BC and 2000 AD.

The new tests carried out in the University of Padua labs were carried out by a number of university professors from various Italian universities and agree that the Shroud dates back to the period when Jesus Christ was crucified in Jerusalem. Final results show that the Shroud fibres examined produced the following dates, all of which are 95% certain and centuries away from the medieval dating obtained with Carbon-14 testing in 1988: the dates given to the Shroud after FT-IR testing, is 300 BC ±400, 200 BC ±500 after Raman testing and 400 AD ±400 after multi-parametric mechanical testing. The average of all three dates is 33 BC ±250 years. The book's authors observed that the uncertainty of this date is less than the single uncertainties and the date is compatible with the historic date of Jesus' death on the cross, which historians claim occurred in 30 AD.

The tests were carried out using tiny fibres of material extracted from the Shroud by micro-analyst Giovanni Riggi di Numana who passed away in 2008 but had participated in the 1988 research project and gave the material to Fanti through the cultural institute Fondazione 3M.[3]

The major criticism of the study concerned the source of the Shroud fibers. The fibers were secured from the estate of Professor Giovanni Riggi di Numana who had been officially deputed to provide the samples for the 1988 Carbon-14 dating. Riggi kept some for later use with the approval, he claimed, of the then Cardinal Archbishop of Turin. Riggi's claims were later disputed by Church authorities although he provided photographs of the additional samples with the Cardinal's seal.

Regardless of the authorization issue, Fanti's study opens the door to new journeys of exploration in the study of the Shroud.

Big Data

Thus far we have considered the Shroud at various levels and from multiple perspectives. Our focus has been on "facts" rather than "ideas." Too many of the discussions on the Shroud are driven by speculation and conjecture rather than by the data. Disagreements on interpretation are reasonable but we should at least start with the data being interpreted.

Here are the facts we have highlighted:

- The image on the Shroud is distinguished by attributes discoverable only with the advent of modern science and technology: it is a photographic negative that is three-dimensional in nature.

- The Man of the Shroud is a Man of Sorrows whose wounds mirror those described in the Passion narratives of the New Testament Gospels.
- There are two kinds of image on the Shroud, the body image and the blood image. There is no body image under the bloodstains.
- Medical professionals, including two of the best-known medical examiners of the modern era, veterans of thousands of autopsies of murder victims, have analyzed the wound and blood patterns found on the Shroud. As clinically experienced doctors, their diagnosis is clear: the victim was scourged, crowned with a caplet of thorns, carrying a heavy object, crucified, pierced on the side with a spear.
- On the basis of their hands-on examination, blood chemistry experts have concluded that the bloodstains on the Shroud are made up of hemoglobin and test positively for serum albumin. In other words, it is real blood.
- Multiple botanical studies of the Shroud have led specialists to conclude that it has pollen grains from plants found only in the Middle East.
- Dust particles on the Shroud have the unique signature of the limestone caves "in and near" Jerusalem.
- The Shroud's style of stitching is similar to that of cloth found in the ruins of the Jewish fortress of Masada dating back to 74 AD.
- The Sudarium of Oviedo, kept in Spain since the eighth century and believed to be the facecloth placed on Jesus after his death, appears to be a counterpart of the Shroud. Studies of the two cloths indicate that there is a match between the facial and neck stains, the nose structure on both and the blood groups.
- As for the Shroud in history, we know this much: ancient accounts speak of an image of Jesus "not made with human hands" as well as of a shroud with his image; iconography apparently inspired by this image resemble the face of the Man

of the Shroud; the ancient image was brought to Constantinople, the capital of the New Rome, in the 10th century and historical descriptions and a surviving Hungarian painting of this image correspond with what we see today on the Shroud (significantly, Constantinople claimed to have the burial linens of Jesus); when the Crusaders sacked Constantinople in the thirteenth century, French knights in particular transported many of its relics to France; in the fourteenth century the Shroud we see today surfaced in a village in France as part of the bequest of a renowned knight killed in battle.

These "facts" still leave us with fundamental questions: how was the image formed and preserved? Why does it have the character of a three-dimensional photographic negative? Why does it stay only on the surface without permeating the underlying linen fibers? And why did the carbon-dating investigators conclude that it is a medieval work?

A distinctive contribution of the present work is its focus on these questions. We have tried to show that the only cogent explanation for the formation of the Shroud image is the action of skin bacteria (e.g., *Staphylococcus epidermidis*) and skin cells on the linen. This microbial action was entirely (and understandably) "missed" by previous Shroud investigators, most memorably by the radiocarbon dating team. Since the latter group failed to decontaminate the Shroud samples at a microbial level, they ended up dating not the Shroud but its latest microbial populations. The different layers of microbial populations, in fact, preserved the Shroud. The Shroud image has the attributes of a photographic negative because of chain reactions involving skin bacteria, oxidation processes and the linen much as photographs were classically created by the oxidation of photosensitive chemicals. The 3D property came from the variation in the density of the microbes found in different parts of the body. The image stays "on the surface" because the image is made up of microbes. All these phenomena can, in fact, be replicated here and now.

The Big Data ensemble of facts outlined here point unmistakably to the conclusion that the Shroud is a burial cloth of a victim of crucifixion from first century Palestine. This is a conclusion accepted by investigators of different religions and none. Die-hard skeptics, of course, remain unfazed since their main concern is not with the facts as we have them but with the ideas they wish to promote. We will study their counter-explanations in more detail in Section IV but before that we should address a more fundamental question: who was the Man of the Shroud? Was it indeed Jesus of Nazareth as we have implicitly assumed?

Section III
Who Was the Man of the Shroud?

In previous sections, we have examined the evidence in favor of the authenticity and antiquity of the Shroud of Turin. We have taken it for granted that, if authentic and ancient, the Shroud is the burial cloth of Jesus. But this assumption needs to be justified. If the Shroud dates back to the first century and was wrapped around the body of a crucified man, how do we know that the Man of the Shroud was Jesus, that its image is an imprint of Jesus' own face and body?

Nature editor Philip Ball, a fair-minded observer, argues that "'authenticity' is not really a scientific issue at all here: even if there were compelling evidence that the shroud was made in first-century Palestine, that would not even come close to establishing that the cloth bears the imprint of Christ."[1] Paradoxically, anti-Shroud polemicist Steven Schafersman holds a different view. He asserts that, "If the shroud is authentic, the image is that of Jesus."[2] In his view, there are only two options: the Shroud was produced either by the body of Jesus or by human artifice. And if it can be shown that the Shroud is an actual burial cloth from the first century then it would have to be the Shroud of Jesus.

Here, however, we will go beyond the low bar set by Schafersman to address Ball's challenge. We have shown that the Shroud is indeed a first century burial cloth. Next, we have to focus on the problem of identifying the Man of the Shroud. How can we possibly know whether or not it is an image of Jesus as traditionally claimed?

Of course, this question is built on a more fundamental assumption: that Jesus existed and was crucified and buried and rose from the dead. Before going further, this assumption needs to be supported with evidence. Did Jesus exist? What evidence is there concerning his purported death by crucifixion and his alleged resurrection from the dead? To this we turn.

3.1 Life, Death and Resurrection

Not even the most ferocious early critics of Christianity argued that its founder did not exist. In the last two centuries, however, a few fringe writers had claimed that Jesus did not exist. But the Jesus-did-not-exist school of thought recently sustained a head wound. Despite three decades of publications denying Jesus' existence, its best-known modern spokesman, G.A. Wells, a British professor of German, admitted in his latest book that a real Galilean Jewish prophet of the early first century named Jesus did indeed exist. Atheist apologist Jeffery Jay Lowder of Internet Infidels reports: "There is simply nothing intrinsically improbable about a historical Jesus; the New Testament alone (or at least portions of it) are reliable enough to provide evidence of a historical Jesus. On this point, it is important to note that even G.A. Wells, who until recently was the champion of the Christ-myth hypothesis, now accepts the historicity of Jesus."[3] Historian Michael Grant points out "no serious scholar has ventured to postulate the non-historicity of Jesus."[4] In fact, no major atheist historian denies the existence of Jesus.

Historical Evidence

We know that Jesus existed on various grounds:

- The extra-biblical evidence available from near-contemporary historians and the undeniable fact that Christianity spread rapidly across the Roman Empire at a very early date.
- The Gospel accounts which even skeptics admit have a historical core.
- The transformation of the Apostles of Jesus after his alleged resurrection from the dead.

We not only find mention of Jesus in extra-biblical Jewish and Roman writings but these references come principally from two individuals who were, respectively, the greatest of the Roman and Jewish historians, Cornelius Tacitus (ca. 56-120) and Flavius Josephus (ca. 37-100).

Furthermore, the explosive growth of churches proclaiming his teachings shortly after the death of Jesus is explicable only if he actually existed. The *Oxford History of the Biblical World* observes that "by the middle of the first century churches had been established in most of the major cities of the empire, including Rome."[5] In other words, within two decades of Jesus' reported death his teaching was proclaimed at the very heart of the Roman Empire.

Our first and most compelling introduction to Jesus comes to us in the Gospel narratives. The portrait presented in the four Gospels show us a personality that can plausibly be considered as not only real but uninventable. It is not just Christians who walk away with this impression. Gandhi, for instance, said: "I do not need either the prophecies or the miracles to establish Jesus' greatness as a teacher. Nothing can be more miraculous than the three years of his ministry."[6]

Another remarkable testimony to the power of the Jesus of the Gospels came from a fellow Jew, Albert Einstein. When asked by G.S. Viereck whether he accepted the historical existence of Jesus, Einstein replied, "Unquestionably. No one can read the Gospels without feeling the actual presence of Jesus. His personality pulsates in every word. No myth is filled with such life. How different, for instance, is the impression which we receive from an account of legendary heroes of antiquity like Theseus. Theseus and other heroes of his type lack the authentic vitality of Jesus."[7]

The Gospels do not just paint a portrait of Jesus. They also claim to be accounts of historical events. This, of course, raises the question of how it is possible to determine their historical accuracy. In the past, New Testament scholars have been notorious for riding off in all directions at once. Three recent developments, however, have resulted in a more rational evaluation of the Gospel narratives.

The Jewish Jesus

Perhaps the greatest breakthrough in biblical studies over the last three decades has been the discovery of the Jewish Jesus. A May 2008 *Time* story described this as one of the "ten ideas that are changing the world." To understand what Jesus said and did and how he was perceived we must understand first the theological thought-world and symbol-universe in which he lived, spoke and acted. He was Jewish and lived in the world of Second Temple Judaism. Now this might seem to be a mere truism but incredibly most of the leading lights in the history of New Testament criticism had seemed entirely oblivious to it. Among other things, eminent scholars over the last two centuries portrayed him as an enemy of Judaism, a purveyor of platitudes, a Sixties hippy and a 19th century liberal German revolutionary. But this has changed with the re-discovery of Jesus' Jewish identity.

The Gospels as Standard Biographies

Another even more recent revolution in New Testament studies has been a new insight into the structure of the Gospels. Rudolf Bultmann was an early 20th century German theologian who set the agenda for New Testament studies in the first half of the century. Many influential scholars of that period accepted the Gospel according to Bultmann: they accepted his view that the literary form of the Gospel was unique – "sui generis" – and each of the four Gospels had been assembled from bits and pieces of oral tradition. In this Bultmannian universe, the evangelists were less writers in their own right than they were collectors of traditions.

In a dramatic turnabout, it is today widely recognized that the sui generis theory was mistaken and the Gospels belong to a biographical genre common in the Greco-Roman world. The Bultmannian quest for the historical Jesus, it is now widely recognized, was simply a wild goose chase.

According to Richard Burridge, author of a pathbreaking work in this area:

- Greco-Roman biographies have internal and external features. Internally, these biographies center on the ancestry, birth, education, character traits, deeds, death and influence of its subjects. From an external standpoint, they tended to be prose narratives with a length of 11-19,000 words and chronological accounts with topical material. In its internal and external features, the structure and content of the Gospels match those of the Greco-Roman biographies.
- The protagonist's last days and death were a key part of the Greco-Roman biographies: Plutarch (17.3%), Nepos (15%), Tacitus (10%) and Philostratus (26%). It is no surprise then that the Last Supper, Trial, Passion, and Resurrection of Jesus takes up between 15 and 19% of the Gospels.[8]

Burridge points out that the very literary structure of the Gospel is a testimony to the Evangelists' central claim about Jesus: "The shift from unconnected anecdotes about Jesus, which resemble rabbinic material, to composing them together in the genre of an ancient biography is not just moving from a Jewish environment to Greco-Roman literature. It is actually making an enormous Christological claim.".

> It is true that "no rabbi is that unique." But the literary form employed, he observes, introduces a new and unexpected dimension: "Writing a biography of Jesus implies the claim that not only is the Torah embodied, but that God himself is uniquely incarnate in this one life, death and resurrection."[9]

This new perspective on the Gospels represents an irreversible breakthrough in our study of Jesus. We realize now that we are dealing with what were seen as standard biographies of Jesus rather than quilts of sayings and traditions.

Historicity

Another breakthrough comes from a new understanding of the Gospels themselves as historical documents. In the past, critics had sought to brush aside the surface credibility of the Gospel narratives by offering their own highly speculative accounts of what (in their view) actually took place. But now, as E.P. Sanders highlighted in his influential *Jesus and Judaism*, the dominant view is that we not only know what Jesus set out to accomplish and what he said but how both of these fit into the world of first century Judaism. Peter Stuhlmacher of Tubingen points out that as a Scripture scholar his inclination was to doubt the Gospel stories but as a historian he is obliged to take them as reliable. In fact, as he sees it, the biblical texts are the best hypotheses for explaining what really happened.

The accounts of the suffering and death of Jesus are especially important. Non-Christian historians accept the crucifixion of Jesus as historical. In a remarkable passage, the Jewish scholar Paula Fredriksen writes, "What do we *know* about Jesus of Nazareth, and how do these facts enable us to start out on the road to a solid and plausible historical portrait of him? The single most solid fact about Jesus' life is his death: he was executed by the Roman prefect Pilate, on or around Passover, in the manner Rome reserved particularly for political insurrectionists, namely, crucifixion."[10] Fredriksen notes that it is unfashionable to talk about "facts" but a reconstruction of Jesus' mission and message must address "what we *know* to have been the case."

There is a new willingness today to pay attention to what is obvious in the texts. Even skeptical historians now acknowledge that the narratives of the Passion in the four gospels, the accounts of the suffering and death of Jesus, are remarkably consistent. They are willing to admit that there is a core of historical truth here. In *A History of Christianity*, Diarmaid MacCulloch points out that "the Passion narratives are probably the earliest continuous material in the Gospels." He notes that the title "King of the Jews" in these narratives "was not a title for which the later Christian Church found any use and so its survival in the tradition is all the more instructive. That 'King of the Jews' phrase is an inescapable repeated refrain through the Passion narratives, even despite the embarrassment which it was to cause Christians."[11] A 2016 Parade story reports, "Were you a first-century charlatan, wishing to create your own religious movement and recruit followers to your cause, this is most assuredly not the story you would create. The Messiah was not, in any version of the Jewish tradition, supposed to be crucified (or subsequently to be raised from the dead). So, yes, we believe that Jesus was crucified."[12]

Resurrection

A central dimension of the phenomenon of Jesus is the electrifying and altogether unexpected event that has come to be called the Resurrection. The claim is that Jesus of Nazareth, the Messiah of Israel, was crucified on a tree and buried in a cave – and then that his body was found to no longer be in the tomb and that he appeared on numerous occasions to his Apostles and other followers, conversing and even eating with them, confirming them in their faith and energizing them to carry on the mission he had entrusted to them.

This dimension of the Rising Son is central because all of Christianity starts with three assumptions:

- Jesus is alive and acting in history;
- he is alive not as an idea but a definite person, an agent who takes specific and concrete actions;
- and we know the first two assumptions to be true because he appeared physically to his disciples after his death.

Traditionally, the historical evidence for the physical resurrection of Jesus has been classified under three categories: the existence of the empty tomb, the reports of the appearance of Jesus to his disciples and others and the origin of the Christian Church. These categories of evidence are summarized below and have been described in detail by William Lane Craig.

Empty Tomb

- Jesus was buried: 1 Corinthians 15; Gospel accounts; no other claims made about Jesus' body.
- Tomb was guarded: even enemies acknowledged this.
- Historical core: sequence of discovery described differently in the gospels but common threads: Joseph of Arimathea buried Jesus; women discovered the tomb was empty; angel seen outside.
- Women shown as first witnesses: Jewish law does not recognize women as witnesses: if the accounts were invented, men would have been the witnesses.
- No fanciful description as in the apocryphal accounts. No description of the actual resurrection.
- Wolfhart Pannenberg, one of Germany's most important 20th century theologians, was once an atheist. He underwent an intellectual conversion to Christianity and affirmed that it is rationally untenable to deny that the resurrection of Christ was a historical event. He considered the empty tomb to be a strong argument in its favor. He noted that it is well known that there were early disputes between Christians and Jews concerning the Resurrection. "The Jews accepted that the tomb was empty. The dispute, however, was about how this is to be explained. The Jews said the disciples had removed the body. But they did not question the fact that the tomb was empty. That, I think, is a very remarkable point. And then, of course, my main reason is a general reflection, given the concreteness of the Jewish understanding of a resurrection from the dead. It would hardly be conceivable that the earliest Christian congregation could have assembled in Jerusalem of all places, where Jesus had died and was buried, if His tomb was intact."[13]

Appearances

- In I Corinthians, Paul reports that Jesus appeared to him, Peter, to the apostles and 500 others. The epistle was written when

many of those who had seen the Risen Jesus were still alive and so it is unlikely to have been concocted. The Gospels show Jesus appearing to Mary Magdalene and other women, to ten of the Apostles and then to all eleven and also to the disciples on the road to Emmaus.
- Did the appearances in fact occur? The legend hypothesis is implausible because I Corinthians 15 is dated very early.
- Were the witnesses simply hallucinating? William Lane Craig gives reasons why this hypothesis does not work: hallucinations are individual in nature and are experienced only by one person but the appearances of the Risen Jesus were witnessed by hundreds; those who hallucinate usually expect to see the object of their hallucination but the terrified disciples did not expect to see their Master again after his crucifixion and burial; hallucinations are sometimes induced by drugs or mental illness but the appearances were witnessed by people with a wide range of personalities and backgrounds.
- Were they lying? This is implausible because the Apostles literally staked their lives on the claim that they had seen the risen Jesus. Almost all of them died horrible deaths – but they went to their deaths proclaiming the resurrection. It is unreasonable to suggest they were willing to die for a lie. Not only did they die proclaiming the resurrection but their lives were transformed as well. Also at least one witness was initially a skeptic, Saul of Tarsus. He had no reason to "invent" the Resurrection since it contradicted his previous beliefs.

Birth of a New World Order

- The explosion of Christianity in first century Palestine cannot be explained without reference to the Resurrection. From the very beginning, as even Bultmann acknowledged, the message that Jesus had risen from the dead lay at the heart of the Christian Gospel. Affirmation that Jesus was the Messiah would have been impossible if he had not been resurrected

because for all practical purposes his mission would be considered a failure if it ended with the cross.
- The claim that Jesus had risen from the grave is not something that could have been extrapolated from any belief-system. Although some Jews believed in the idea of resurrection, such a resurrection is one that involved the entire human race and that took place at the end of history. Jesus' resurrection, however, involved one individual and it took place within the historical process.

Even the hardiest of today's skeptics acknowledge the fact that the earliest Christians had experienced something. What remains in dispute is the nature of that "something."

By far the most compelling feature of the claim of resurrection is the transformation of the followers of Jesus and the genesis of the Christian movement. What transformed eleven fearful peasants and fisher-folk into superheroes who preached the Good News across the world despite trials and tribulations and eventually horrendous deaths? What galvanized them to take on the most powerful empire of the day? The hallucination hypothesis is not remotely plausible for those who know the causes and characteristics of hallucinations (an individual experience, drugs, mental illness, an expectation of seeing that which is allegedly witnessed – none of which apply in the present case). The idea of a hoax is wildly implausible given the improbability that anyone would embrace a gruesome end in order to perpetrate a fiction. And the notion that a transplanted myth was the impetus behind the conversion and martyrdom of a ragtag band of one-time cowards and whiners stands self-condemned simply in the telling. It seems undeniable that the extraordinary transformation could only be explained by an extraordinary event and, in this respect, the Resurrection makes perfect sense.

Two noted Jewish scholars have made this precise case.

The first is Sholem Asch, perhaps the leading Jewish writer of his generation: in 1936 he was named among the world's ten greatest living Jews along with Einstein and Freud, the only writer on the list. Asch held that "What must remain an eternal mystery to those who are blind and deaf enough not to believe in miracles is the spread of Christianity during the first three hundred years. No intellectual evidence, no rationalistic explanation can clarify the phenomenon or see it as anything other than an extraordinary development which remains outside the bounds of our intellectual, sensible point of view."[14]

A more detailed defense comes from the Jewish author Pinchas Lapide. In rejecting the idea that the Resurrection appearances were the result of autosuggestion, he writes, "In none of the cases where rabbinic literature speaks of such visions [springing from autosuggestion] did it result in an essential change in the life of the resuscitated or of those who had experienced the visions. Only the vision remains which was retold in believing wonderment and sometimes also embellished, but it did not have any noticeable consequences. It is different with the disciples of Jesus on that Easter Sunday. Despite all the legendary embellishments, in the oldest records there remains a recognizable historical kernel which cannot simply be demythologized. When this scared, frightened band of the apostles which was just about to throw away everything in order to flee in despair to Galilee; when these peasants, shepherds, and fishermen, who betrayed and denied their master and then failed him miserably, suddenly could be changed overnight into a confident mission society, convinced of salvation and able to work with much more success after Easter than before Easter, then no vision or hallucination is sufficient to explain such a revolutionary transformation. For a sect or school or an order, perhaps a single vision would have been sufficient – but not for a world religion which was able to conquer the Occident thanks to the Easter faith. ... If the defeated and depressed group of disciples overnight could change into a victorious movement of faith, based only on autosuggestion or self-deception – without a fundamental faith experience – then this would be a much greater miracle than the resurrection itself."[15]

About Bultmann and other reductionist explanations, Lapide writes, "Most of these and similar conceptions strike me as all too abstract and scholarly to explain the fact that the solid hillbillies from Galilee who, for the very real reason of the crucifixion of their master, were saddened to death, were changed within a short period of time into a jubilant community of believers. Such a post-Easter change, which was no less real than sudden and unexpected, certainly needed a concrete foundation which can by no means exclude the possibility of any physical resurrection. One thing we may assume with certainty: neither the Twelve nor the early church believed in the ingenious wisdom of theologians!"[16]

Lapide concludes, "I cannot rid myself of the impression that some modern Christian theologians are ashamed of the material facticity of the resurrection. Their varying attempts at dehistoricizing the Easter experience which give the lie to the four evangelists are simply not understandable to me in any other way. Indeed, the four authors of the Gospels definitely compete with one another in illustrating the tangible, substantial dimension of this resurrection explicitly.

"Often it seems as if renowned New Testament scholars in our days want to insert a kind of ideological or dogmatic curtain between the pre-Easter and the risen Jesus in order to protect the latter against any kind of contamination by earthly three-dimensionality. However, for the first Christians who thought, believed, and hoped in a Jewish manner, the immediate historicity was not only a part of that happening but the indispensable precondition for the recognition of its significance for salvation."[17]

Riding off into the SonRise

Despite the skepticism of past generations, recent times have seen an intellectual resurrection of the Resurrection story. Five monumental works have helped drive this dramatic turnabout:

- "The Events of Easter and the Empty Tomb" (1952) by Hans von Campenhausen which defended the historical basis of the empty tomb claim.
- *Jesus – God and Man* (1968) by German scholar Wolfhart Pannenberg. He argued that from the evidence for the empty tomb and the appearances of Jesus it is possible to rationally conclude that Jesus' resurrection from the dead took place in

history. As we have seen, Pannenberg was an atheist whose conversion was occasioned by his study of the Resurrection.

- *The Resurrection of Jesus – A Jewish Perspective* (1983) by Pinchas Lapide, the source of our previous citations.
- *The Resurrection of the Son of God* (2003) an epochal work by N.T. Wright that tied together the historical arguments in favor of the bodily resurrection of Christ. Wright notes that the question to be answered is "Why did Christianity emerge so rapidly, with such power, and why did believers risk everything to teach that Jesus really rose?" The best explanation, in his view, is that "Jesus' tomb was discovered empty on Easter morning" and "Jesus then appeared to his followers alive in bodily form."[18]
- *The Resurrection of God Incarnate* (also 2003) by noted Oxford philosopher Richard Swinburne. Swinburne argued that his resurrection from the dead is what you would expect if Jesus is God incarnate, that the accounts of his appearances were such as would be expected if there were real appearances and that the available evidence is what would be expected if the tomb was indeed empty.

Clearly then the claim of the resurrection of Jesus can no longer be dismissed as a pious legend or an afterthought. Of course, the empty tomb, the posthumous appearances of Jesus and the transformation and worldwide witness of the apostles do not "prove" the resurrection. The fundamental starting point has to be the very origin of the Christian phenomenon. N.T. Wright argues that "The empty tomb and the meetings with Jesus are, in combination, the only possible explanation for the stories and beliefs that grew up so quickly among his followers. ... I have examined elsewhere all the alternative explanations, ancient and modern, for the rise of the early church, and I have to say that far and away the best historical explanation is that Jesus of Nazareth, having been thoroughly dead and buried, really was raised to life on the third day with a new *kind* of physical body which left an empty tomb behind it because it had 'used up' the material of Jesus' original body, and which possessed new properties which nobody had expected or imagined but which generated significant mutations in the thinking of those who encountered it. If something like this happened, it would perfectly explain why Christianity began and why it took the shape it did."[19]

There is of course much rational evidence but by itself such evidence does not suffice because the claim it tries to support is too extraordinary to rest simply on a structure of premises and inferences. No one put this more poignantly than the most influential philosopher of the 20th century, Ludwig Wittgenstein, another modern Jewish witness to the Resurrection: "What inclines even me to believe in Christ's resurrection? It is as though I play with the thought. If he did not rise from the dead, then he is decomposed in the grave like any other man. HE IS DEAD AND DECOMPOSED. In that case he is a teacher like any other and can no longer HELP; and once more we are orphaned and alone. So we have to content ourselves with wisdom and speculation. We are in a sort of hell where we can do nothing but dream, roofed in, as it were, and cut off from heaven. But if I am to be REALLY saved, - what I need is CERTAINTY - not wisdom, dreams or speculations - and this certainty is faith. And faith is what is needed by my HEART, my SOUL, not my speculative intelligence. For it is my soul with its passions, as it were with its flesh and blood, that has to be saved, not my abstract mind. Perhaps we can say: Only LOVE can believe the Resurrection. Or: It is LOVE that believes the Resurrection."[20]

3.2 Connecting the Dots

We have seen that Jesus was indeed a historic person, that his crucifixion took place in recorded history and that his resurrection from the dead seems to be the only cogent explanation for otherwise inexplicable phenomena. We have also seen that the Shroud of Turin was indeed the burial cloth of a man who was scourged, crowned with thorns, crucified and pierced on the side.

Now we must address the questions with which we began: Who was the man of the Shroud? Was this the Shroud of the crucified Jesus? Is its image the image of Jesus?

The only viable approach in addressing these questions is that of the cumulative case: of showing how a multitude of dots paint a picture. This is, in fact, the standard modus operandi in dealing with historical questions. It is how historians build a life of Julius Caesar and map the exploits of Alexander the Great.

A series of standalone "facts" (documents, monuments, coins, et. al.) morph into a full-blown narrative. Diverse data-points form a pattern. But someone has to connect the dots, see the big picture. This is an act of judgment, the triggering of an insight, a leap to the light.

In many instances, there is a consensus on the emergent end-result. Yet on questions like the existence of a non-physical dimension of the human person or the identity of Jesus, there are sharp disagreements even among scholars. But these kinds of issues are too important to be left to experts and specialists. In fact, in certain cases, ideological commitments or inflexible habits of thought can blind the "very elect" to the very obvious. And just as the jury system of determining innocence or guilt relies on the judgment of ordinary men and women, so also the interpretation of hard facts is a domain that belongs to "the least of these."

As we will see in the next section, the arguments of some Shroud foes demonstrate the truth of Cicero's warning: "there's no opinion so absurd that a philosopher hasn't expressed it." But right now we are concerned with the problem of who was the man of the Shroud. And in addressing this question we have no choice but to let the facts speak for "themselves" rather than yielding the floor to fact-free theories and agenda-driven spin. Our focus is on fact-finding and fact-checking so as to grasp the truth underlying the facts.

Let us start with the facts of the matter that have already been reviewed in some detail:

- The New Testament narratives speak of the deceased Jesus being wrapped in a linen shroud and laid to rest in a tomb. "Taking the body, Joseph wrapped it [in] clean linen and laid it in his new tomb that he had hewn in the rock." (*Matthew* 27:59-60).
- The narratives conspicuously tell us that the discovery of the Resurrection of Jesus was also a discovery of the linen shroud that had wrapped his body. "But Peter got up and ran to the tomb, bent down, and saw the burial cloths alone." (*Luke* 24:12).
- Apocryphal works in the early centuries of Christianity refer to the existence of a shroud with Jesus' image.
- Edessa was reputed to be the home of this image icon until it was moved to Constantinople in 944 – the "image not made by man of Christ our God was transferred to Constantinople."
- Writings of the time state that the shroud of Christ is stored in the royal treasury of Constantinople and displayed on rare occasions. The 1192 Hungarian Pray Manuscript, a depiction of this image in Constantinople, bears a striking resemblance to the Shroud of Turin.
- After the 1204 sack of Constantinople by the Crusaders, the image disappears. According to one report it was taken to Athens and then to France.
- In 1357, the cloth now known as the Shroud of Turin is displayed by a French knight with the claim that it is the shroud

of Christ. The cloth showed faint images of the face and torso of a crucified man. The cloth was passed on to the ruling family of Italy who bequeathed it to the Catholic Church in the 20th century.

- In 1898, when the Shroud is first photographed, it turns out that the Shroud itself is like a negative film with its positive picture emerging when photographed. This picture matches the description of the Passion narratives. The Man of the Shroud is seen to have been scourged, crowned with thorns, crucified and speared with the Roman instruments of scourging and crucifixion exactly as described in the Gospels. There are wounds on the wrists, feet, hands and side accompanied by the expected blood flow patterns. On his head, there are wound marks consistent with a caplet of thorns. He has about one hundred whip marks and bruises on the face, knees and shoulders, all of which are consistent with the accounts of being scourged, struck on the face, carrying the cross and falling under its weight.
- The facial and neck stains on the Shroud match the bloodstains on the Sudarium of Oviedo which is believed to have been the facecloth placed on Jesus after his death. This facecloth was removed before the body was wrapped in the shroud.
- That the Shroud exists is consistent with the Gospel claims of the Resurrection. Through the action of digestive enyzmes in the soil, burial shrouds decay and disintegrate simultaneously with the bodies they wrap. But if Jesus' body "disappeared," then it would not have played a role in the disintegration of its shroud.
- Studies show that the bloodstains on the Shroud have such chemical properties of blood as hemoglobin and serum albumin.
- The Shroud has on it pollen grains and floral images consistent with an origin in Israel-Palestine.
- The Shroud shows traces of the limestone dust found in the ancient tombs in Jerusalem.

- The 21st century breakthroughs in microbiology led to an especially significant finding: not only did skin bacteria from the Man of the Shroud produce the image on the Shroud *but the image as we have it could have formed only if he went through the sequence of actions depicted in the Passion narratives within the timeframe recorded there.* According to the narratives, the period between the prosecution of Jesus and his execution was six to eight hours. This was the minimal time period required for the unrestricted growth of the skin bacterium *Staphylococcus epidermidis* so that the oxidation products from the excess bacteria could become visible on the linen surface (greater than 1 million bacteria/cm^2 of skin surface). If Jesus had been put to death immediately there would have been no image formed other than the blood marks. Likewise, without the scourging and head wounds caused by the crown of thorns and the continuous bleeding, there would be no image. If Jesus had died during the scourging and implantation of the crown of thorns, there would be no image. If Jesus had died carrying his cross, there would be no image. Time was essential to generate the number of bacteria needed to be transferred from the deceased body to the surface of the burial linen to produce the image. It was a matter of all or nothing.

Now if Jesus of Nazareth really existed and was crucified as described in the Gospels, and we have good reason to affirm both, then it cannot be denied that the Shroud of Turin is at the very least a plausible candidate for being considered his burial shroud. This is apparent simply from the match between the state of the body of the Man of the Shroud and the particular wounds said to have been inflicted on Jesus. This surface plausibility is further strengthened by the geographical exactitude of the pollen and the sand found on the Shroud. Further, we face the extraordinary fact that this is a shroud without a body – a fact that ties in with the Gospel accounts of Jesus' resurrection. If he rose from the dead, it is no surprise that there is no body in the Shroud. In fact, if there had been a body, the shroud would have decayed along with the body. If the body had been physically removed, such displacement would have left its mark on the Shroud. "If the shroud had been lifted off the man, one of two things would have happened: If the blood was still wet the stain on the cloth would smear; if the blood was dry it would have broken the crusted blood that had soaked into the weave. Neither occurred, thus leading some researchers to believe that the body must somehow have dematerialized without the removal of the shroud. If the shroud merely collapsed and was not thrown back, then the story of Peter and John's arrival at the tomb after Jesus' Resurrection (John 20:1–10) makes better sense when Peter saw 'the linen cloths lying' and John 'saw and believed.'"[21]

Drilling deeper with respect to the "missing body," here is what we know:

1. The image on the Shroud does not show a decomposing body.
2. The Shroud was found without any body or skeletal remains.
3. Normally shrouds are not found because they decompose with the bodies wrapped in them.
4. So the survival of the Shroud suggests that the body was missing.
5. Furthermore, a scenario in which the body was moved out of the Shroud is implausible because such "removal" would have left markings of some kind on the Shroud. The blood clots on it would have been smeared or broken if there was any moving of the body but this is not the case. In his 1995 book *The Double Images on the Shroud* (*La doppie immagini della Sacra Sindone*), the Italian neurosurgeon Nicolo Cinquemani has pointed out that any movement of the body would have resulted in breakage of the clots and tearing of the threads. Since there is no evidence of either, the body was not removed. In fact, anyone removing the body would have removed the shroud in which it was wrapped as well.

On a natural level, then, the Shroud bears witness to the dematerialization of the body it enveloped. Burial shrouds and human remains decompose simultaneously in the soil where it is acted on by an array of digestive enzymes. Microorganisms that produce the enzyme cellulase are mainly associated with soils. These are responsible for the decomposition of cellulose-containing materials such as roots and stalks of vegetative crops and other plant material including man-made linens. Various enzymes engineer the decomposition of human tissues. Now if the body in the Shroud did not decompose, if it "disappeared," then it is no surprise that the Shroud survived. As we have seen, the body was not physically removed since any such action would have left a bloody trail. And since the body in the Shroud image shows no sign of decomposition, it is clear that it "left" relatively soon after being placed in the tomb.

In brief, the Man of the Shroud was both crucified and resurrected. We do not HOW he rose from the dead. We just know THAT he did indeed rise from the dead. And of no other person has this claim been made other than Jesus of Nazareth who "was crucified, died and was buried" and "rose again from the dead."

So, in identifying the Man of the Shroud with Jesus, we move from surface plausibility to supporting evidence to the singularity represented by the resurrection. The clincher, the smoking gun, is the 21st century revelation of the role of the microbiome: only a person who underwent precisely the tortures and injuries described in the Gospels in the specific timeline found there could have "created" the image on the Shroud. No other scenario can account for the coming to be of the particular image we see there as also of its photographic and three dimensional properties. Thousands of people were crucified in the first century and many thousands more were scourged. But there is no account of anyone being crowned with thorns other than Jesus who was mocked as the "king of the Jews." The image of the Man of the Shroud then bears the unique signature of the crucified Jesus: struck on the face, scourged, crowned with thorns, made to bear the beams of the cross on his shoulders, crucified, pierced on the side, wrapped in a shroud, laid to rest in a tomb – over a period of six to eight hours. This signature coupled with the resurrection singularity built into the Shroud tells us that the Man of the Shroud was indeed Jesus of Nazareth.

The correspondence between the Man of the Shroud and the Jesus of the Gospels has been usefully summarized by various writers.

Herbert Thurston who was, incidentally, a fervent critic of the Shroud:

> As to the identity of the body whose image is seen on the Shroud, no question is possible. The five wounds, the cruel flagellation, the punctures encircling the head, can still be clearly distinguished. If this is not the impression of the Christ, it was designed as the counterfeit of that impression. In no other person since the world began could these details be verified.[22]

John Iannone:

> The study of the testimony of the Gospels, when matched with Roman weapons and practices of crucifixion as well as with the findings of medical pathologists studying the Shroud, shows a strong correlation of these sources. The uniqueness of the markings on the Shroud, especially when taken in their totality, with the testimony of the Gospels provides the signature or the fingerprint of the Crucifixion that identifies the Man of the Shroud with Jesus. This is especially true of the capping of thorns to mock Jesus' 'kingship,' a unique event never recorded with any other crucifixion victim; the lancing of the right side to assure that Jesus was dead instead of the usual crucifragium; the nails through the wrists; the scourge marks all over his back; the marks of the crossbeam (patibulum) on the shoulders, and the swollen face from the beating of the Sanhedrin guards. In all cases, the words match the wounds which match the weapons.[23]

Robert Spitzer, S.J., "Science and the Shroud of Turin":

> Why would we think the body in the Shroud was that of Jesus? As explained above, it is exceedingly improbable that the Shroud is a medieval forgery. First, there are no paints, dyes or

other pigments on the Shroud (except for the small flecks coming from the sanctification of icons and paintings which touched it). Secondly, the anatomical precision of the blood stains—which are real human blood that congealed on the Shroud before the formation of the image—are in precise anatomical correlation to the image itself. How could a medieval forger have accomplished this? Thirdly, it is exceedingly difficult to explain how pollen grains indigenous to Palestine appeared in abundance on a shroud of probable Semitic origin (if it originated in medieval Europe) ... How could a medieval forger have duplicated these first century Palestinian characteristics of the Shroud?

He notes that

1. The material of the Shroud, the pollen grains on it, have their origin in First Century Palestine – the place where Jesus was purported to have died.
2. The blood stains come from a crucifixion event identical to the one described in the four Gospels – which was very unusual, if not unique, in many respects – such as being crowned with thorns, being flogged, and being pierced with a Roman pilium.

Spitzer concludes that "The odds of this First Century Palestinian burial shroud – with the unique features of Jesus' crucifixion and resurrection – being that of anyone else is exceedingly remote."[24]

By assembling the "hard facts" relating to the Shroud and comparing them to the Gospel accounts of the passion, death and resurrection of Jesus, we have created a comprehensive case. A big picture emerges from this connecting of the dots: the Man of the Shroud is none other than the Jesus we meet in the four Gospels. What is more, the Shroud itself is the fifth Gospel!

Section IV
Shroud in a Shroud

Some theories are so "preposterously silly that only very learned men could have thought of them. But such theories are frequently countenanced by the naive since they are put forward in highly technical terms by learned persons who are themselves too confused to know exactly what they mean." [1]
Philosopher C.D. Broad

"I really do believe that our attitudes are shaped much more by our social groups than they are by facts on the ground. We are not great reasoners. Most people don't like to think at all, or like to think as little as possible. And by most, I mean roughly 70 percent of the population. Even the rest seem to devote a lot of their resources to justifying beliefs that they want to hold, as opposed to forming credible beliefs based only on fact. Think about if you were to utter a fact that contradicted the opinions of the majority of those in your social group. You pay a price for that. ... The decisions we make, the attitudes we form, the judgments we make, depend very much on what other people are thinking. ... One danger is that if I think I understand because the people around me think they understand, and the people around me all think they understand because the people around them all think they understand, then it turns out we can all have this strong sense of understanding even though no one really has any idea what they're talking about. [2]
Cognitive scientist Steven Sloman

"One has to belong to the intelligentsia to believe things like that. No ordinary man could be such a fool." [3]
George Orwell

4.1 Guilty Until Proven Guilty

We have already seen that the usual arguments against the Shroud's authenticity are demonstrably false: the Carbon 14-based dating, the McCrone Mutiny, the doctored D'Arcis memorandum, the medieval prop thesis.

But as each counter-argument goes down, every few months, it would seem, there is a new theory "explaining" the origin of the Shroud. Each seeks to show that the shroud had nothing to do with the Jesus of the New Testament. In fact, it was "as everyone knows" a product of the Middle Ages. Incredibly enough, a medieval mastermind pioneers the science of photography using the rudiments of the day (although the use of this technology would not be detected for centuries). Throw in 3D as part of "special effects." And sprinkle pollen and a few grains of sand from the Middle East for a realistic feel (again detectable only in later centuries). Alternately we can always take refuge in the pious skeptical tradition of the Shroud being an ingenious painting. True, there are no brush-strokes, no directionality, no penetration of the fibres, no pigments. No way to coordinate the blood and body image positioning. No way to explain why the image did not follow the art conventions of the day. Nor the theological taboos (the representation of an unclothed Jesus). And, of course, the photographic negative and 3D effects "just happened."

Plausibility matters little to the skeptics' "and then a miracle happened" school of thought. The only objective is to stop the barbarians at the gate – the hordes of fundamentalist fanatics and crazed clerics eager to burnish and brandish this weapon of mass instruction. We should use all the means at hand, cruel and unusual though they may be, to keep the knights of faith at bay. Coincidences galore, spectacular feats of speculation, ideas long before their time. At all costs, the show must go on.

But no matter how entertaining all this may be, at some point we have to leave the circus tent. There's a real world out there where the tigers are not jumping through hoops and acrobats are not leaping from one trapeze to the next. Checkbooks have to be balanced, the laws of gravity have to be obeyed, tumors have to be detected and detached. The real world operates with rules of evidence and explanation. It assumes above all the reign of sanity. Flights of fancy, silly seasons, freak shows – these are barnacles hanging onto a ship not the ship itself. So it is with the inquiry into the origin of the Shroud's image.

With this we reach the day of reckoning for the origin theories. We cannot continue expending time and energy on theories that have long since lost all right to be heard. If a theory has a fatal flaw, there is no point chasing it down to the bitter end. We simply dispatch it with an obituary and even an eulogy if especially colorful. It is basically a whack-a-mole expedition, a matter of clearing the weeds. In that spirit, we outline the most popular and exotic origin theories with the specific mortal wounds that put them out of their misery. No more and no less.

It should be noted that a common characteristic of all the "alternate" theories of Shroud origin is their desperate nature. This is because desperate times call for desperate measures. So, no matter how seemingly artificial, how quixotic it may be, each theory has the saving grace of serving as an escape route from the obvious. But, as a result, we are left with a bunch of chickens with their heads cut off running off in every direction at once. Admittedly two of the theories (the vaporographic and Maillard reaction theories) do not deny that the Shroud was the burial cloth of Jesus, but their proposed mechanisms of action are, like the others, both wildly implausible and fatally flawed.

4.2 And then a Miracle Happened! Unnatural Shroud Origin Theories

Vaporware

The Shroud image was caused by vapors/gases emanating from the body (e.g., ammonia or lactic acid from sweat). This theory was first introduced by Paul Vignon. There have been many variations since.

Here are the problems. Vapors "penetrate and permeate" cloth. A vapor image would necessarily be blurry because gases simply diffuse. And it could not be 3D or include an image of hair or display the bloodstains as currently observed. For all these and other reasons, it could not be the mechanism of action for the Shroud image. This image is 3D, strictly on the surface and not diffused within the cloth, sharp when seen from a distance and includes hair. Finally, the chemicals associated with such vaporagraphic transfer were not found on the Shroud.

Another suggestion is that the image could have been produced via the Volckringer effect but, if feasible, this would take years whereas the corpse was in the Shroud only for a short time. This hypothesis is not plausible since it cannot be replicated and similar images are not found on any other shroud.

It has also been claimed that the image was a Kirlian aura. Theories of esoteric energies like Kirlian auras fall outside the purview of replicable science and are therefore irrelevant to an evidence-based inquiry. "The Kirlian aura will remain a fascination to *non-scientific* people ... with words like 'life-force,' 'photic energy,' 'bioplasma,' ... and so on. Most Kirlian claims will come from 'experimenters' who will combine the complicated effects of gaseous discharges with samples having complicated structures and electrical properties, and film recordings involving complicated photographic processes and interpretations based on ignorance of the phenomena and the need for proper controls."[4]

Hot Air

Then there is the Malliard Reaction theory. According to the theory as proposed by Raymond Rogers, the body of Jesus while supposedly decomposing gaseously emitted amino acids. These reacted with the carbohydrates on the Shroud, changed the coloration of its fiber and gave us the image we have.

But there is no sign of any kind of decomposition of the body on the Shroud. Also, for the reaction to take place, the temperature of the body would have to have been about 104F. Given the rigor mortis of the body wrapped in the Shroud, this would have been impossible. Another anomaly: this kind of reaction would have resulted in different kinds of coloration but there is uniform coloration across the Shroud. And, of course, the reaction would not give us the negative and 3D effects that are distinctive features of the Shroud image. Most importantly, if this kind of reaction is possible, why is it that no other shrouds have carried body images.

Here's the rub

One of the more bizarre origin theories, proposed in the Seventies by Joe Nickell, an amateur magician, is the "rubbing" theory. A wet cloth was wrapped around a bas relief sculpture of a corpse (bas-relief is a technique where the figure is not more than a few inches higher than the flat background). Iron oxide/rouge powder used by jewellers was applied to the cloth once it dried and the image was formed. When STURP showed that the Shroud had at best a few flecks of iron oxide, Nickell explained this by saying that the pigment had just fallen off.

The problems here are manifold. If a pigment was used as hypothesized it would penetrate the cloth which it obviously has not. Most important, Nickell's own *twentieth century* attempts to replicate the rubbing technique have produced blurry images and not the sharp clarity and resolution or the 3D effect of the Shroud image. And if produced in the Middle Ages, the portrait would have nothing to do with the art conventions of the time. William Meacham points out that "there is no rubbing from the entire medieval period that is even remotely comparable to the Shroud, nor is there any negative painting. Nickell's wet-mold-dry-daub technique was not known in medieval times ... and even that technique fails to reproduce the contour precision and three-dimensional effect, the lack of saturation points, and the resolution of the Shroud image."[5] De Wesselow says, "Nickell implies an art-historical episode so bizarre, speculative, impractical and anachronistic that it is quite unbelievable."[6]

Dead on Arrival

More sophisticated but no less desperate is the model proposed by Nicholas Allen, a South African art historian. In his re-creation, a medieval inventor "happened upon" the techniques and raw material required for photography and the Shroud image is the result. Half a millennium before the discovery of photography, the mystery inventor rubbed light-sensitive silver nitrate/sulphate onto a cloth kept in a dark room with an aperture with a quartz lens on the wall facing it. A cast of a body was placed in front of the aperture so as to face the cloth. When exposed to sunlight over a few days the image supposedly appeared.

But here are the flies in the emulsion. No records, no reports, exist of so momentous a discovery. The kind of quartz lens required for the hypothesis to get off the ground was available only in the 19th century. No traces of silver nitrate were found in the hands-on investigation of the Shroud. Images produced with this method are significantly different from the Shroud image in terms of light directionality.

Cruel and Unusual

Other recent attempts usually recycle the older theories. For instance, in an initiative funded by an association of Italian atheists (yes, they do exist), the chemist Luigi Garlaschelli sought to show in 2009 that the Shroud image could have been made with medieval material. A volunteer was wrapped in a linen cloth that was "rubbed" with a red ocher pigment. A bas relief mask was used as the face; sulfuric acid, etc. were applied to the cloth. To artificially create the impression of aging, the cloth was heated and then washed. Blood and water stains and holes were added later. Garlaschelli said the pigment would have faded away over the ages leaving only the ghostly image.

The Achilles heel of this theory, as of most of the fake relic theories, is the mystery of the missing pigments. It's all very well to say that it faded over the ages but inevitably, if painted, the pigments would have penetrated the fibres and not stayed at the surface. In fact, even with the passage of time, there would still be thousands of particles in the image fibres. None were found. STURP ruled out all known paints, dyes and pigments. Moreover, if there were pigments, these would have been affected by the fires survived by the Shroud not to speak of the water that fell on it and the rolling and bending of the cloth through the centuries. No such discoloration and variation is to be found on the Shroud. In fact, unlike the Shroud, there is no continuity of coloration on Garlaschelli's image. Also, red ochre in the Middle Ages had various impurities such as manganese and none of these have been found on the Shroud. What little iron oxide there is on the Shroud came from the linen retting process and from other paintings known to have been touched to the Shroud: and these are found in image and non-image areas. Studies showed that the blood and serum stains on the Shroud preceded the image itself. Garlaschelli, however, added his liquid tempera "blood" onto the image. STURP researcher John Jackson observed that, from a 3D perspective, Garlaschelli's image is "really rather grotesque. The hands are embedded into the body and the legs have unnatural looking lumps and bumps."[7]

Over-shadowed

Shroud image creation is an equal opportunity employer. English instructor Nathan Wilson entered the fray and offered his own Top Model theory. A forger painted the Shroud image on a large piece of glass, placed it on an unbleached linen cloth and left it to bleach in the sun for days. After the linen lightened, the painting on the glass formed the darker side of the cloth with an image on it (the painting acts as a mask preventing bleaching action on the linen under it). Talk of seeing through a glass, darkly!

After its 15 seconds of fame, this Shadow Shroud theory has itself (appropriately) faded into the shadows. Its first and most egregious error was its anachronism: the precise kind of plate glass required for the proposed image creation was available not in the Middle Ages but in the nineteenth century. Flat plate glass itself was made only in the seventeenth century. But even if the glass were miraculously procured by a medieval forger, other problems are terminal. As with all the other "fake" theories, the image portion on the Shadow Shroud (unlike the real Shroud) is not superficial: it has the same color underneath as it has on the top. Also, after all these centuries, there is no bleaching action (by sunlight or chemicals) on the Shroud image: if it were unbleached linen as the Shadow theory claims this would not be the case. Again, there is no image under the bloodstains on the Shroud, something unaddressed by the theory.

Bloodless Coup

As we have seen, a favorite trope of the Shroud-debunking school is the use of methods and mechanisms that re-create features of the Shroud in such a fashion as to refute the traditional claim that it was the burial cloth of Jesus. Predictably, these attempts generate sensational headlines for a standard news cycle and then disappear without a trace once subjected to critical scrutiny.

Aother recent example of this phenomenon is a headline in the *Journal of Forensic Science* about the "false bloodstains" on the Shroud. According to the two Italian scientists who led this study, forensic anthropologist Matteo Borrini and the redoubtable Luigi Garlaschelli mentioned earlier, Bloodstain Pattern Analysis (BPA) of the stains on the Shroud indicates that the shape and direction of the blood flows found there are inconsistent with a crucified corpse.

They admit that the blood flow from top to bottom on the Shroud is consistent with a crucified body since the blood is driven by gravity from the chest to the bottom. But they contend that blood flow in the back across the lumbar region (as found on the Shroud) is not possible for a crucified body. If the flow had happened after the corpse was laid in the Shroud, then the blood should have flowed from the chest to the armpits and then behind the shoulders. Instead you find blood in the chest region but also stains in the lumbar region that go horizontally from left to right. The basis for their argument is an experiment performed with a mannequin and a volunteer.

Although published only in 2018, the study reported in the paper had been carried out in 2014. Both blood flow analysts and Shroud experts find the report's assumptions and resultant conclusions to be implausible and, in fact, irrelevant. The study was fatally flawed in both the methodology used and the assumptions that led to the conclusions.

As pointed out in 2014, the blood used for the study included an anti-coagulant which made it different from the blood expected from a tortured and crucified victim:

> A crucified victim's "blood had to be more viscous than normal and therefore the pathways of the streams coming out of the wounds may have taken very different directions from those of the fluidized blood used in this experiment. Another parameter that influences the path of the dripping is the speed at which the blood comes out of the wounds of the man of the Shroud, which is unknown, therefore it is not possible to reproduce it in an experiment like that of Borrini and Garlaschelli".

> Secondly, "The skin of the crucified one had to be sweated, dirty with soil, swollen with hematomas and encrusted with blood from the whip-inflicted wounds. In short, anything but the smooth, clean skin used in the experiment. And it is precisely the state of the skin, the incrustations, the swellings, the dirt, the sweat that may have interfered in an important way on the direction of the drippings of that thick and viscous blood."

With regard to the blood in the lumbar region, "We don't know if the blood spillage from the wound to the side can be simulated realistically (same speed, same scope) by squeezing out a blood-soaked sponge. We do not know if the Shroud was used only to wrap the corpse of the crucifix or also to transport it from the cross to the tomb: in which case, taking the body by the arms and the feet, the part of the basin would have found itself pushed down and lower, causing blood stagnation at the height of the belt. We are in the field of pure hypothesis."[8]

It should be noted that past studies done with the corpses of individuals who died of the same causes as those believed to have led to the death of the Man of the Shroud support the blood flow seen on the Shroud. These are more likely to be accurate than one done with a mannequin.

BPA authority "Jonathyn Priest said the study did not take into account if someone cleaned, carried or prepared the body to be buried."[9]

Finally, "Victor Weedn, chairman of forensic sciences at George Washington University in Washington, D.C., said ... he was "skeptical of this analysis," saying there was no reason to believe that the body could not have been moved while being transported. "We're not dealing with things we really know about," Weedn said. "We just don't know if this cloth was laid on someone who just laid there or was wrapped around the body or moved some before being put in a particular place."[10]

Last and least

For comic relief, we finally turn to the Da Vinci past-life theory proffered by paranormal/occult/alien-hunter authors Lynn Picknett and Clive Prince. Approximately one hundred years *before* he was born near the village of Anciano, Italy, in 1452, Leonardo Da Vinci created the Shroud as a photographic self-portrait. It was a fourteen-foot photo; the head being his own and the body belonging to a cadaver he was dissecting. His objective was to have generations to come bow down before him. The evidence? He was a genius and there is no reason why he couldn't have created a proto-camera and taken a proto-photograph. Did he say he had performed such a feat? No, but he had his reasons (unknown to us) for keeping silent on the matter. Does the theory have any problems? Yes, the Shroud was known to have been displayed in Europe by 1355-7, nearly one hundred years before Da Vinci was born. Let's leave it at that.

Here's the final nail in the coffin for all the medieval painting theories. How can you create a "fake" shroud in the Middle Ages when every modern attempt to create a fake has failed to replicate the fundamental features of the actual Shroud? There is no way around this unless we join those skeptics who believe in medieval miracles.

4.3 Mistakes were Made ...

A review of these intellectual somersaults and backflips can leave us bewildered. What lies behind these efforts to dogmatically deny the most obvious explanation for the image on the Shroud? Here is an image of a dead man whose wounds correspond with those of the Jesus of the Gospels and whose visage matches replicas of the Image of Edessa. What is more the image has unique features discoverable only with the tools of modern science and explicable only if it wrapped an actual body. Then there is the pollen from Palestine. Given all this, there is no reason why we should not consider the possibility that the Shroud might indeed be the burial cloth of Jesus. It is clearly not a question of faith since Jews, agnostics and non-Christians of various kinds have had no problem in reaching this conclusion.

But the very idea of such a connection has people of a certain temperament running for the exits. They would rather miraculously bring Da Vinci to life a hundred years before he was actually born than consider the raw data testifying to the Shroud's non-miraculous connection to a historical person. They would rather invoke a medieval deus ex machina than admit the possibility that the Shroud is a shroud. They fervently urge us to believe in non-belief so that we can understand how to not-understand. Faith of this variety admits of no argument. What is required at this point is not scientific or historical analysis but psychoanalysis.

Independent observers who accept the results of the Carbon 14 studies nevertheless dismiss the medieval miracle theories. A remarkable case in point is Michael Tite who supervised the 1988 carbon dating. Tite, who was Director of the Oxford University Research Laboratory for Archaeology and the History of Art and editor of the journal *Archaeometry*, said in a BBC interview that the

Shroud was a shroud and not a painting. "There's no real evidence for paint. I don't think so [it was painted]. The other odd thing is if you look at every painting from the Middle Ages and later the Renaissance, they always paint Christ with the nails going through the palm of the hand and through the top of the foot. Whereas, in reality, If you're going to crucify someone in order that they stay on the cross, you have to put the nails through the wrist and through the ankles [as is the case in the Shroud]. I think complete replication of the image hasn't been successfully achieved. I don't believe it's the Shroud but I think it's highly probable there was a body in there."[11] In another interview, he observed: "It is a remarkable image. I never looked at it the way I look at paintings of Christ."[12]

Similarly, *Nature* editor Philip Ball notes, "It [the Shroud image] does not seem to have been painted, at least with any known historical pigments. ... Among the flaky theories about the shroud's origin is one that it was created by Leonardo da Vinci, using a primitive photographic technique to record his own image. You couldn't make it up (although people do). The photographic hypothesis has been developed (so to speak) in some detail, notably by South African art historian Nicholas Allen. He has even used medieval materials to create faint photographic images on linen cloth saturated with silver nitrate. But Allen failed to convince other shroud scholars, who reasonably asked how an invention as marvellous as photography could have remained otherwise unknown until the nineteenth century."[13]

On the positive side, archaeolologist William Meacham lays out what we do know:

> There are many more flaws in the "powerful case" for medieval artifice, and I must beg the reader's forbearance for what must begin to seem like the whipping of a very dead horse. There is no medieval depiction of scourge marks of such realism (radiation and fine detail) or correspondence to the Roman scourge. The nude figure of Christ is extremely rare, unheard of in an object for public veneration, and Shroud copyists generally saw fit to correct it. The wrist-nailing is

unique, according to art historian McNair (1978:35): "I have studied hundreds of paintings, sculptures and carvings of Christ's crucifixion and deposition, from the 13th to the 16th centuries, and not one of them shows a nail wound anywhere but in the palm of the hand." Depiction of a non-circlet crown of thorns severing the head is extremely rare. The Shroud is unlike any 14th-century or earlier artist's conception of the deposition and wrapping in linen. The portrayal of the face is extremely close to the Byzantine style, as Whanger has shown. It is clear, therefore, that clever artistry simply cannot be stretched to cover such a wide range of extraordinary circumstance. Innovation, even at genius level, is bounded by the cultural context and cannot diverge therefrom to the extent that the Shroud contradicts the 14th-century milieu. From this massive conflict between the Shroud and medieval art 1 believe there can be only one conclusion - that the Shroud image belongs to the 1st millennium, with the corollary that it is the imprint of a body. These conclusions should now be considered well-documented archaeological judgements, approaching the level of certainty if normal standards are applied, especially since they agree exactly with the evidence from medical studies.

I have not invented this historical knowledge (dating and assessment) or failed to present it, as Schafersman and Pellicori maintain. The identification and dating of artifacts by their cultural affinities is part and parcel of archaeology. A web of intricate, interlocking, field-tested evidence is usually taken as proof, though not exactly comparable to proof in the natural sciences, as Maloney and Otterbein point out. With the medical, pollen, blood, pigment, and art historical evidence all pointing away from medieval forgery and collectively indicating the Shroud's origin in the ancient Middle East, the issue of Stage 1 authentication should have been settled after the archaeological confirmation of three cultural traits first hypothesized from Shroud studies - the Roman wrist-nailing ...

and supine, hands-over-pelvis burial posture. The skeptics, however, posit such miraculous qualities in the "clever artist" that, by the same criteria, no artifact, manuscript, or work of art could ever be dated or authenticated. And contrary to Alcock, science and history do proceed by decisions of validity and authenticity. ...

The dating, geographical origin, and association with Christ are indicated not by an isolated feature or datum, but by a web of intricate, corroborating detail as specific as that used in the authentication of a manuscript or painting and certainly as reliable as many other archaeological/historical identifications which are generally accepted.[14]

With the advent of the Internet, every conceivable opinion or outlandish theory has suddenly gained credibility and immortality. If we wish to evaluate the Shroud on its own terms, we need to step away from this madding crowd and separate facts from speculation. The endless discussions about the Shroud's authenticity are symptomatic of a besetting sin of the modern mind: the preference for speculation over hard facts.

In all the skeptical theories, inferential speculation is preferred over observational science. It is Galileo all over again: what the ordinary observer sees through the camera or microsope or spectrometer is denied by the Spin Doctors of the Skeptical Church. It might be argued that the C14 study involved observational science but even here we are dealing with inference rather than direct observation. The results of what was directly observed involved here are not in doubt. Rather, it is *the failure to fully observe* that is the issue: the failure to measure and monitor a crucially relevant datum: the microbial contamination that radically skewed the resultant inference.

But what about the "definitive" dating of the cloth in 1988? Let us first understand what it is not:

- It is not a scientific observation of a fact comparable to an astronomical observation of a distant star.

- It is not a scientific theory built around immediately experienced events as was the case when the apple hit Newton and led him to formulate the theory of gravity.

Rather, even if the microbial contamination had been addressed,

- It is an educated guess akin to the occasional claim of having "discovered" life on Mars based on one tenuous piece of apparent data and large servings of speculation.

But it is now undeniable that the investigators failed to microbially decontaminate their samples and so their inferences are fatally flawed. On the positive side, it is undeniable too that the paradigm of the microbiome shows us how the image was created and then preserved over twenty centuries.

Inevitably, the preference for agenda-driven speculation over inconvenient fact has driven discussion of the Shroud. We have before us an image on a piece of fabric that mirrors in exquisite detail the crucified Jesus of the Gospels. The image is demonstrably not a painting and its origin and continued existence on the fabric are simply not explicable by modern – let alone medieval – science. That it is a burial cloth conforming to the description of Jesus' burial cloth in the New Testament is indisputable. That it bears similarities to the first millennium Image of Edessa and the Shroud of Constantinople is undeniable. Its connection to the Sudarium of Oviedo, the pollen from Palestine, et al – all these are the hard facts before us HERE and NOW.

Our central focus should be on the image on the Shroud visible now in a way impossible before the 19th century. Whose image is it and how did it come to be? All hypotheses and attempted explanations should be driven by these two questions. But this is not how the modern (already made up) "mind" works. The experts must be called in and their verdicts are all that matters. As always, their analysis will start and end with speculation. The historians say it is a medieval artifact because relics were commonly "created" in the Middle Ages, etc. (of course, they can point to nothing comparable to the Shroud). The scientists say we will date the fabric with the latest age-measurement technologies and based on these tests they *infer* that it is medieval in origin. The experts have spoken. The case is closed. (Although, as we have seen, these inferences are demonstrably mistaken).

All moderners now believe that the Shroud is a clever medieval creation. This is as obvious a truth as the universe being eternal and our minds simply being matter in motion. Thus is born a new superstition to be uncritically accepted by anyone who wants to be accepted in polite "modern" company.

But, as is their wont, facts do not go away simply because we cannot face them. The haunting image. Its whole infrastructure of "supporting evidence." The paradigm of the microbiome. The failure to produce any comprehensive or coherent theory of medieval forgery (without recourse to miracles). These continue to cry out for an explanation. The skeptics we will always have with us. They have no barriers to entry, no accountability for mindlessly propagating egregious errors. But for those who wish to remain faithful to the facts and to be led by the evidence as we have it, the explanation of it all is obvious and undeniable. There can be no reasonable doubt that the Shroud is an artifact from the first century that can best be explained as the burial cloth of Jesus.

Section V
Shroud Unshrouded

5.1 Seeing is Believing

A persistent theme of this book is the importance of focusing on the facts before us. Principal among these is the artiFACT itself: the Shroud with its image. The "facts" in the context of the Shroud are *observational* facts. What we observe, what we perceive, is the touchstone of truth in this inquiry.

Seeing is believing.

The primary observational facts form a Big Data infrastructure:

- A linen cloth with bloodstains and a blurry image of a wounded man seen with the naked eye.
- A sharp and startling image of the same man revealed to the naked eye when a photograph of the cloth is developed indicating that the image has the properties of a photographic negative.
- A three-dimensional bas relief of the man's image that emerges when it is fed into the VP-8 image analyzer (used by NASA to decode three-dimensional data in images of planets and stars) indicating that the image contains 3-D data and that the cloth was wrapped around a body.
- Fluorescence microscopes, optical and scanning electron microscopes, et. al. verify the presence of blood on the cloth.
- Scanning Electron Microscopy and Energy Dispersive Spectrometry identify geographically specific pollen grains on the cloth which include grains from the Middle East.
- Doctors' diagnoses clinically authenticate an identity of wounds and bloodflows between the Jesus of the Gospel Passion narratives and the man imaged on the cloth.

- Ancient icons of Christ and the illustration of Jesus being laid on a shroud in the 12th century Hungarian Pray Codex "match" the image on the cloth.
- Application of skin bacteria to the faces of human subjects that are then covered with cloth results in the creation of facial images with attributes similar to the image under study (negative, 3D, etc.).

All of these are observational facts: instances of "seeing". All have been investigated and "certified" by specialists in the respective domains under study (e.g., botanists analyzing the pollen grains). Individually and collectively they testify to the conclusion that the cloth under study was an authentic burial cloth. Additionally, specific characteristics of the image ranging from the caplet of thorns and the wound on the side to the fact that the body in the cloth did not decompose but "disappeared" indicate that this was the burial cloth of the resurrected Jesus of Nazareth referenced in the New Testament accounts.

Cumulative Case

The collective "testimony" of the observational facts constitutes a cumulative case. And in this volume we have presented a cumulative case for the affirmation that the Shroud of Turin is the burial cloth of Jesus of Nazareth. But what is meant by "cumulative case" and can it show something to be true? And how does it apply to the Shroud?

By "cumulative" we mean an accumulation of different data points. Equally applicable is this dictionary definition: "growing in quality, strength or effect by successive additions or gradual steps."[1] A cumulative case is built on "converging and convincing" probabilities. But what can a cumulative case show? To what extent can we trust its conclusions?

What most people may not realize is that most of our beliefs and choices (including life and death decisions) are based on "cumulative cases" of various types. On the natural level, we have two ways of knowing: through the senses and through our conceptual analysis of sensory data. The latter depends on our ability to construct a cumulative case. When we argue a legal case, we are presenting a conclusion derived from a number of "facts" and conjectures. When we propose a scientific theory, we are offering a framework that "explains" a kaleidoscope of observations. When we speak of a historical event, we rely on a vast or meager rolodex of testimonies, artifacts and extrapolations. The acts of mapping, aggregating and integrating all involve creating or drawing on a cumulative case of some kind. Everyday life would grind to a halt if we cannot trust our cumulative case for the existence of a world outside us which follows certain laws or if we refuse to act without a scientific demonstration supporting every third party statement. Snap judgments and global interpretations are the stuff of life.

Cumulative cases, then, are not just normal but indispensable. But how do we know if such a case has "proved its point?" Unanimity of acceptance cannot be a criterion because reasonably intelligent and informed individuals differ on fairly basic claims. But there are certain other criteria that must be met for a "cumulative case" to even receive a hearing:

- A powerful body of supporting evidence
- The ability to explain both surface and deeper level data
- Reasoning processes that are generally deemed plausible

We contend that the case for the Shroud's authenticity laid out here not only meets these criteria but that the preponderance of evidence points beyond a reasonable doubt to its being the burial cloth of Christ.

A cumulative case can be complemented by the absence of a plausible counter-explanation. This applies in the case for the Shroud. The Shroud skeptics' counter-explanations are considerably handicapped by the absence of serious evidence, inability to explain either the surface or underlying data and, finally, the wholly implausible and wildly speculative nature of the reasoning patterns deployed. The skeptical arguments tend to also contradict each other. For instance, one critic says the image on the Shroud is not realistic enough to be genuine, another says it is too realistic to be genuine. It's a matter of throwing everything against a wall and hoping something will stick. Even worse, the skeptics' case depends on a series of medieval and other miracles (e.g., producing a photograph half a millennium before the discovery of photography).

They have chosen the path of believing without seeing.

Serious students of the Shroud, on the other hand, have always chosen to let the evidence set the agenda. Over a century ago, the French agnostic scientist Yves Delage made a cumulative case argument for the Shroud:

> I willingly recognize that none of these given arguments offer the features of an irrefutable demonstration, but it must be recognized that their whole *constitutes a bundle of imposing probabilities*, some of which are very near to being proven. A religious question has been needlessly injected into a problem which in itself is purely scientific with the result that feelings have run high and and reason has been led astray. If, instead of Christ, there were a question of some person like a Sargon, an Achilles or one of the Pharaohs, no one would have thought of making any objection. In dealing with this question, I have been faithful to the true spirit of science, intent only on the truth, not concerned in the least whether or not it might impinge on the interests of any religious group. I recognize Christ as a historical personage and I see no reason why anyone should be scandalized that there still exist material traces of his earthly life."[2]

More recently, STURP scientist William Meacham argued that the case for the Shroud's authenticity is as strong as that for other historical artifacts:

> In each of the examples of historical "fact" which I have compared with the Shroud – Tutankhamen's tomb, the dating of the Parthenon, Shakespeare's authorship, Hitler's death, the Lascaux paintings, the Shang dynasty – there is an element of

the circumstantial, and nothing irrefutable, but careful investigation of the pattern of interlocking data, unique features, and the extraordinary circumstance required by alternative explanations leaves no reasonable doubt, and no substantial reason to doubt, unless one has a particular axe to grind. So it is with the Shroud. ... And, as Burridge points out, so many of our supposed certainties, even in the natural sciences, are actually possibilities or probabilities that we need to be continuously reminded of the frailty and lack of absolute certainty inherent in our knowledge.[3]

Shroud photographer Barrie Schwortz put it well when he said: "Really, it's an accumulation of thousands of little tiny bits of evidence that, when put together, are overwhelming in favor of its authenticity."[4]

The cumulative case for the Shroud is unique in one respect: it is modern science that has enabled us to discover the Shroud for what it truly is. Without photography, fluorescence microscopes, and various other technologies, we could not have unraveled the hidden secrets of the Shroud. It is science, then, that is our primary supporting tool in studying this artifact. Not science as a vehicle for speculative inference but science as a tool of observation here and now. The kind of science that yields hard perceptible facts. The rubric of science extends only to what can be quantified: to the "measuring" of physical objects and processes. Anything beyond that is inference: sometimes verifiable (through further experiments, independent replication, etc.), sometimes speculative but reasonable and sometimes simply a flight of fancy. Quantifying is one action, inferring the other.

Mapping the Microbiome

The scientific dimension that has been particularly emphasized here is the microbial one. It is a novel and necessary way of "seeing" any ancient artifact and as such represents an irrevocable advance in Shroud studies. No one denies that burial cloths have microbes attached to them but, given the state of science in previous eras, the nature and significance of these microbes was not sufficiently considered. Inevitable though it was, this impoverishing "oversight" can no longer be left unaddressed in an era when the impact of the microbial world is better known than ever before.

Microbes (bacteria, fungi, viruses and most protozoa), of course, are of too small a size to be seen by the unaided human eye. They must be observed by microscopes that can greatly magnify their dimensions. This "invisibility" hampered their discovery until an amateur scientist Anton van Leeuwenhoek invented crude microscopes and began showing drawings of the "animalcules" in 1673. For this discovery he is referred to as the Father of Microbiology. As microscopes and their magnifications improved over the years, their immense impact in disease, role in natural recycling of organic matter, production of unlimited products for human use have become quite apparent. They essentially can be found on every object and surface on the earth as well as in the atmosphere.

We have seen that microbes (bacteria) have been here for almost three quarters of the entirety of the earth's existence. Compare that time span to the relatively short history of humanity (*Homo sapiens sapiens*) – a brief moment in the history of the earth. Bacteria are responsible for the appearance of oxygen on the earth allowing the subsequent appearance of plants and animals. It can be safely said that the earth would be an uninhabitable planet without the presence of bacteria.

What microbiology has learned through the modern tools of DNA technology and molecular biology is that we have only begun to "scratch the surface." We now know that we have only identified and cultured about 1% of the microbes that are present. This number will probably be revised lower as we discover new sites for occupation of microbes. Keep in mind that they have been at this business of life eons longer than other more modern life forms. In addition to the process of growth and reproduction, some bacteria have developed life forms (spores) that can retain the essential features of life but not grow and reproduce for millions of years.

The Human Microbiome Project seeks to identify all microbial life forms in specific areas of the human body through the use of advanced methods in molecular biology and DNA technology. With this information, it will be possible to compare disease states with normal healthy conditions. Projects of the most interest include the gastrointestinal tract, the oral cavity including teeth, gingival areas and mucosal surfaces, and the skin the largest organ of the human body. It is estimated that the skin supports trillions of microbes and any two individuals share only 13% of their total bacteria. The remaining 87% vary in such detail that they can be used much like fingerprints to identify specific individuals and follow their location in different environments.[5]

Human skin is composed of three major layers, the outer epidermis, the middle dermis, and the inner hypodermis. In terms of using microbes to identify specific individuals, the outer epidermis is the most important for identification purposes. Multiple layers of dead epidermal cells are sloughed off continuously as the skin replenishes itself. Attached to these outer layers are bacteria that remain with the released skin cells. Movement of the body in walking and other daily activities promotes the exfoliation of bits of outer skin. Thus, a trail of exfoliated skin is found wherever we are located and on all objects that come in contact with these bacterial-coated skin tissues.

Using DNA-based technology, specific bacteria associated with given individuals can be identified.[6] Uses for such technology are almost endless. Even without fingerprints, specific locations of subjects at crime scenes as well as on objects and victims would be conclusive. The identification of multiple birth babies or identical twins could be accomplished since only 13% of the bacteria are identical on given individuals. Arguments over the authenticity of works of art or historical objects or documents could be answered by comparing known personal objects of the artist with the work in question. This would be accomplished with no damage to the piece of art or historical artifact. Refinements in the technology where multiple individuals touch an object still need to be accomplished and will likely involve advanced statistical analysis.

When considering the relatively benign environment of the Shroud of Turin, we should bear in mind the development of the Shroud microbiome on and throughout the ancient linen surface and the vast opportunities for microorganisms to make it their home in unlimited numbers.

The distinctive contribution of this book is its attention to the microbial dimension of the Shroud. This dimension tells us how the Shroud image was formed and preserved. It also shows us why dating studies are fatally flawed if they fail to decontaminate at a microbial level – and why effective microbial decontamination would not be of much better help because it would result in the destruction of the sample to be tested.

Nobody in 1988 (when the radiocarbon dating study was done) grasped the magnitude of microbiology. Now virtually everyone knows at least something about it: the reality of the microbiome, the billions of bacteria in a probiotic supplement, the fact that everything is covered with microbes: this is now as obvious as relativity and quantum physics. When we apply this "glimpse of the obvious" to the Shroud, we can say with certainty that it would have been coated with billions of bacteria. We have no way to know all the details of its microbial history in the centuries following its use as a burial cloth but we do know that it has layers upon layers of bacteria which have an impact on its weight, survival, color, etc. It's a certainty too that its image was formed by bacteria that are now dead. And that it was itself preserved by bacteria. And for the first time we have replicable scientific methodologies that demonstrate the mechanisms responsible for the formation of the Shroud image.

When it comes to the Shroud, as always, it is science that has served as ally and aide in unveiling the momentousness of its mystery. This is the headline that has been obscured in recent years but that now, in the interest of truth and sanity, needs to be retrieved.

Endgame

The cumulative case laid out in this book allows us to reach our own definitive conclusions about the Shroud. But this means we have to look at the totality of the data. It also means we have to connect all kinds of dots – raw data, hard facts, fundamental insights and, finally, circumstantial evidence congruent with what is already known. In connecting the "dots" and explaining what seems evident, we remember that here we have both evidence and explanation as our starting point: the explanation (it is a shroud!) having been handed down centuries ago and the evidence becoming increasingly visible (literally!) over the centuries. To make progress, we have to start with the hard facts and let these drive our inquiry instead of starting and staying with the speculative. And the only plausible way to come to a rational resolution is to consider the cumulative case – for or against authenticity – in the light of all the relevant data points.

Here we have looked at *all* of the relevant data with particular emphasis on the hard facts. Our conclusion is that they comprise a cumulative case that presents us with a big picture:

The Shroud of Turin is at one level the burial garment of Jesus of Nazareth and at another a real-time embodiment of his Passion and a testament to his resurrection from the dead.

5.2 Here and Now – Suffering and the Shroud

"Our hearts remain full of unseen idols until we are stretched on the wood of the Cross with Christ, until we cease trying to nourish ourselves and our desires, and give ourselves completely to the poor, to the needy, to the suffering members of Christ's body throughout the world ... in the presence of God-become-man – stripped and naked, scourged and covered with spittle, dying unto death for love – and man seeking to become like God, raised above his ugliness and misery through love, finding in This Man all his joy, all his love, all the meaning of life and history."

Francois Mauriac, Anguish and Joy of the Christian Life.

As the burial cloth of Jesus, the Shroud of Turin is one of humanity's most precious possessions.

On a scientific level, it is a phenomenon without parallel. It is the microbial creation of the first photograph in history – a three-dimensional one at that. Amazingly, this process could took place only because of the incredibly precise fine-tuning of timing and circumstance involved – the specific time frames involved in the sequence of scourging, crucifixion, death, shrouding and burial. Skeptics have hurled the latest modern technologies at the Shroud seeking to expose it as a fraud only to find that even more advanced technologies have validated its authenticity and unveiled unsuspected new depths to its dynamism. And with each new discovery, the Shroud seems almost to tell us: what took you so long?

From a historical standpoint, it brings home to us the factual underpinning of the Christian story. Jesus of Nazareth actually lived! He suffered, died and rose again. This is the history lesson taught by the Shroud for the nearly 2000 years it has been with us (and let us not forget that in those two millennia it seemed always one step ahead of its own destruction, e.g., at the hands of Roman persecutors, Christian and Islamic Iconoclasts, the French Revolution). While New Testament scholars ponder the Gospel texts and hypothesize and speculate and infer and guess, the Shroud appears on the horizon of human history as a stubborn hard fact that cannot be explained away. It tells the Gospel truth!

On a theological plane, we are stunned by the raw reality of Jesus' suffering. Sobering, numbing, bewildering. Not only did Jesus die but he died a horrifying death. This real-time portrait of pain tells us more than any text, preacher or movie. It is a picture that tells a thousand stories. The literal flesh and blood of the fallen Messiah is crying out to us. We cannot turn our eyes away from those haunting eyes closed in death. The scourging, the piercing, the crowning with thorns, the crucifying, all burst out of the cloth into our very being. If we have hearts they should break at this requiem for the Redeemer, if we have eyes we must weep for unto us was born this Savior who was crucified for us. Only in encountering the suffering Jesus can we bear our own suffering. For the Shroud is nothing less than an elegy from eternity inviting us to the glory of the Third Day.

End Notes

* Shroud of Turin image source-
https://commons.wikimedia.org/wiki/File:
Full_length_negatives_of_the_shroud_of_Turin.jpg

I Shroud 3.0

1.1 The Once and Future Shroud
1. http://bit.ly/2szNDP4.
2. http://go.nature.com/2rUnKrk.

1.2 Shroud 2.0 Hard Facts
1. Barbara J. Culliton, "The Mystery of the Shroud of Turin Challenges 20th-Century Science," *Science*, Vol. 201, 21 July 1978, pp.235-239.
2. http://bit.ly/2sUxB4Z.
3. http://bit.ly/2tMKqLE.
4. http://bit.ly/2sAKoXS.
5. http://bit.ly/2tsaYT5.

1.4 The Jesus Microbiome Project
1. Ian Wilson, *The Shroud* (London: Bantam Books, 2010), 136.
2. https://www.enviromedica.com/learn/microbiome-101-understanding-gut-microbiota/.
3. https://www.sciencealert.com/how-many-bacteria-cells-outnumber-human-cells-microbiome-science.
4. "Microbial Odor Profile of Polyester and Cotton Clothes after a Fitness Session," Chris Callewaert, et al, *Applied and Environmental Microbiology* 80:6611-6619 (2014).
5. http://go.nature.com/2oHZO8Q.

1.5 The Embedded Microbiome that Created the First Photograph in History

1. http://go.nature.com/2rR5i7M.

1.7 It Seemed Like a Good Idea at the Time

1. http://go.nature.com/2sTL6Cn.
2. http://bit.ly/2rVxWj7.
3. http://nyti.ms/2sOFetm.
4. Willy Woelfli, "Archaeological Sherd Dating: Comparison of TL Techniques with Radiocarbon Dates by Beta Counting and Accelerator Techniques," International Radiocarbon Conference, Trondheim, Norway, 1985.
5. Willy Woelfli, "Advances in Accelerator Mass Spectrometry", *Nuclear Instruments and Methods in Physics Research*, B29 (1987), 5.
6. http://bit.ly/2soXeKh.
7. Tyrer, J., *British Society for the Turin Shroud Newsletter*, 20 October 1988, p.11.
8. http://bit.ly/2sOUZAg.
9. Wilson, I. & Schwortz, B., *The Turin Shroud: The Illustrated Evidence* (London: Michael O'Mara Books, 2000), pp.98-100.
10. Thomas De Wesselow, *The Sign* (New York: Dutton, 2012), 163.
11. The Shroud of Turin: A Critical Summary of Observations, Data and Hypotheses, 92. http://bit.ly/2rRuKdm.
12. De Wesselow, op cit., 163-164.
13. http://www.ncregister.com/daily-news/the-shroud-of-turin-latest-study-deepens-mystery.
14. https://www.hommenouveau.fr/2900/religion/des-chercheurs-remettent-en-cause-l-idee-selon-laquelle-brle-saint-suaire-daterait-du-moyen-age.htm?fbclid=IwAR3UfvSEf1F0M5RFc_NMhbE12gCrr2xxD3u-mbEof7BMpBukMlMBCOz1t2U.
15. https://www.acistampa.com/story/sindone-ecco-perche-la-datazione-e-da-rifare-11793.
16. https://www.acistampa.com/story/la-sindone-e-il-grande-flop-della-datazione-medioevale-la-parola-agli-studiosi-11792.
17. De Wesselow, op cit., 167-8.

18 Ibid., 168-9.
19 Ibid., 171-2.

1.10 Letting Sleeping Bugs Lie
1. G.K. Best and S.J. Mattingly, "Chemical analysis of cell walls and autolytic digests *of Bacillus psychrophilus,*" *Journal of Bacteriology,* Vol. 115: 221–227 (1973).
2. http://go.nature.com/2sTL6Cn.
3. http://bit.ly/2rTXRaV.

1.11 FAQs – Critiques of the Microbial Paradigm and Responses
1. http://bit.ly/2szfIG6.
2. http://bit.ly/2rPcBwJ.
3. http://bit.ly/2rPoiU2.
4. http://bit.ly/2soIdb6.
5. TLC-TV: *In Pursuit of the Shroud,* Dec. 22, 1998. Cited in Konikiewicz, L.W., "Turin Shroud and the Science: Digital Enhancement Provides New Evidence" (Chicago, IL: Panorama Publishing: 1999), 44-45.
6. http://bit.ly/2rRuKdm.
7. http://bit.ly/2rPoiU2.
8. http://bit.ly/2rRNCJg.
9. http://bit.ly/2sSwfrI.
10. http://bit.ly/2sOLnG2.

II Shroud In Toto

2.1 Based on a True Story
1. William Meacham, "The Authentication of the Turin Shroud: An Issue in Archaeological Epistemology," *Current Anthropology,* Vol. 24, No. 3 (June 1983). http://bit.ly/2sTUaHk.
2. http://cbsn.ws/2sztxof.
3. http://bit.ly/2osnxLV.
4. http://bit.ly/2oa5FHb.
5. Maurus Green, "Enshrouded in silence. In search of the First Millennium of the Holy Shroud," *The Ampleforth Journal* 3 (1969), 321-345. http://bit.ly/2soITgE.

⁶ Kenneth E. Stevenson and Gary R. Habermas, *Verdict on the Shroud* (Ann Arbor, MI: Servant, 1981), 48-9.
⁷ Ibid., 46.

2.2 Ecce Homo
1. http://bit.ly/2pfmPl3.
2. http://bit.ly/2p2aP9J.
3. http://bit.ly/2szAGVm.
4. http://bit.ly/2pddyNs.
5. http://bit.ly/2odLMPm.
6. http://bit.ly/2rBKSLX.
7. Raymond E. Brown, *The Death of the Messiah*, (New York: Doubleday, 1994, 2 volumes), 206.
8. Ibid., 1210.
9. De Wesselow, op cit., 112.
10. Ibid., 132-3.
11. Ibid., 135.
12. Ibid., 136.
13. http://bit.ly/2pfmPl3.
14. De Wesselow, op cit., 145.
15. Ibid., 147.
16. http://bit.ly/2oxu2yN.
17. http://bit.ly/2rTVcxZ.
18. De Wesselow, op cit., 148.

2.3 Doctors' Diagnoses
1. William D. Edwards, et al., "On the Physical Death of Jesus Christ," *Journal of the American Medical Association*, Vol 255, No. 11, March 21, 1986, 1455-1463.
2. Ibid., 1455.
3. Ibid., 1456.
4. Ibid., 1457.
5. Ibid., 1458.
6. Ibid., 1459.
7. Ibid., 1459.
8. Ibid., 1459.

[9] Ibid., 1460.
[10] Ibid., 1460.
[11] Ibid., 1460.
[12] Ibid., 1461.
[13] Ibid., 1462.
[14] Ibid., 1463.
[15] Ibid., 1463.
[16] Ibid., 1464.
[17] http://bit.ly/2pddAoc.
[18] http://bit.ly/2oxu2yN.
[19] http://bit.ly/2sSsH96.
[20] http://bit.ly/2oxnos0.
[21] http://bit.ly/2pA9ZgQ.
[22] http://bit.ly/1wADJL6.
[23] http://bit.ly/2szvvEZ.
[24] http://bit.ly/2pde8ui.
[25] https://www.lifesitenews.com/news/new-discoveries-prove-man-on-shroud-of-turin-was-really-crucified

2.4 Bloodstains

[1] Schwalbe, L.A. & Rogers, R.N., "Physics and Chemistry of the Shroud of Turin: Summary of the 1978 Investigation," *Analytica Chimica Acta*, Vol. 135, 1982, 3-49.
[2] John H. Heller and Alan D. Adler, "A Chemical Investigation of the Shroud of Turin," *Canadian Society of Forensic Sciences Journal*, 14 (3), 1981, 52.
[3] Heller, J.H., *Report on the Shroud of Turin* (Boston MA: Houghton Mifflin Co, 1983), 215-216.
[4] Vittorio Guerrera, *The Shroud of Turin – A Case for Authenticity*, (Charlotte, North Carolina: Tan 2013).
[5] De Wesselow, op cit., 135-7 ff.
[6] Harry Gove, *Relic, Icon or Hoax?* (CRC Press: Boca Rotan, FL, 1996), 19.
[7] Ibid., 19.
[8] Ibid., 190.

9   John Ray, "Not made by hands," *Times Literary Supplement*, October 19, 1996.
10 http://bit.ly/2obddXz.
11 C. Bernard Ruffin, *The Shroud of Turin* (Huntington, Indiana: Our Sunday Visitor, 1999), 97.
12 http://bit.ly/2obddXz.
13 http://bit.ly/2rTQ90Q.
14 http://bit.ly/2osz4uG.
15 http://bit.ly/2tX8oEX.

2.5 "The stones will cry out"
1   http://bit.ly/2osD0eK.
2   Ibid.
3   http://go.nature.com/2szFz0W.
4   http://onlinelibrary.wiley.com/doi/10.1111/arcm.12269/abstract.
5   http://bit.ly/2ts3eQM.
6   *Archaeological Discovery*, 2015, 3, 158-178,. http://bit.ly/2soUi0c.
7   Danin, Avinoam, Whanger, Alan D., Baruch, Uri, Whanger, Mary. *Flora of the Shroud of Turin* (St. Louis, MO; Missouri Botanical Garden Press, 1999).
8   http://bit.ly/2tMKqLE.
9   C. Bernard Ruffin, op cit..103.
10 Wilson, I., *The Blood and the Shroud: New Evidence that the World's Most Sacred Relic is Real* (New York NY: Simon & Schuster:, 1998), 104-106.
11 http://bit.ly/2pdrSWa.
12 http://bit.ly/2oG4aB3.
13 http://bit.ly/2oegkNH.
14 http://bit.ly/2oxl6cG.
15 http://bit.ly/2sP3dbW.
16 https://www.catholicnewsagency.com/news/scientist_says_to_be_very_careful_when_interpreting_writing_on_shroud_of_turin.
17 https://www.messengersaintanthony.com/content/new-light-shroud.
18 https://www.catholicnewsagency.com/news/scientist_says_to_be_very_careful_when_interpreting_writing_on_shroud_of_turin.

19. https://www.shroud.com/pdfs/guscin2.pdf.
20. https://www.cbsnews.com/news/does-text-prove-shroud-of-turin-is-real/.
21. https://www.telegraph.co.uk/news/religion/6617018/Jesus-Christs-death-certificate-found-on-Turin-Shroud.html.

2.6 Sudarium and Shroud – a Tale of Two Images
1. http://bit.ly/2oxGBd9.
2. Guscin, M., *The Oviedo Cloth* (Cambridge UK: Lutterworth Press, 1998), 30,32.
3. Ibid., 28.
4. EDP Sciences, 2015, http://bit.ly/2pfzOmY.
5. http://bit.ly/2pfzILS.
6. http://bit.ly/2pzXYYK.

2.7 The Shroud – A History
1. http://bit.ly/2ts3MGk.
2. Martha C. Howell and Walter Prevenier, *From Reliable Sources: An Introduction to Historical Methods* (New York: Cornell University Press, 2001), 73-74.
3. De Wesselow, op cit., 184.
4. Scavone, D.C., *The Shroud of Turin: Opposing Viewpoints* (San Diego, CA: Greenhaven Press, 1989), 74.
5. Ibid., 76-77.
6. Cogswell, T.B., *Reconciliation of the Shroud with the Gospels* (Newton, Ma: 1939), 3-5.
7. St. Ambrose, **On the Mysteries**, (Nn. 52-54. 58: SC 25 bis, 186-188. 190).
8. St. Basil the Great, **On the Holy Spirit**, 27.
9. http://bit.ly/2rXoKLk.
10. Pietro Savio, Sindonological Prospectus, 1982, http://bit.ly/2sSsrac.
11. http://bit.ly/2sCod3b.
12. http://bit.ly/2soITgE.
13. Margery Wardrop, *Life of Nino* (Oxford: Clarendon Press, 1803),11.

14. M. GREEN, "Enshrouded in silence. In search of the First Millennium of the Holy Shroud," *The Ampleforth Journal*, Vol. 74, Part 3 (1969), 321-345. http://bit.ly/2soITgE.
15. Ibid., 329.
16. Guscin, Mark, *The Image of Edessa* (Leiden & Boston: Brill, 2009), 151-152.
17. Cited in John C. Iannone, *The Mystery of the Shroud of Turin* (Staten Island, New York: Alba House, 1998), 110.
18. "On the Ill-treatment (Lat.: *Passio*) of the Image of our Lord Jesus Christ at Beirut." published in the *Patrologia Graeca* at PG 28.818 B. Also, *The Sermon of Athanasius*, Mansi documents of the Councils (Mansi 13.384A). Also in the modern collection of Greek "hagiography": BHG 780-788b. References kindly provided by Fr. Brian Daley, S.J.
19. http://bit.ly/2sSsrac.
20. Foreword and Supplement by Hugh Schonfield in K. de Proszynski, *The Authentic Photograph of Christ* (London: 1932), 7 (http://bit.ly/2soITgE).
21. Ibid., 54f.
22. William Meacham, "The Authentication of the Turin Shroud: An Issue in Archaeological Epistemology," op cit.
23. Emanuela Marinelli, "The Shroud and the iconography of Christ," http://bit.ly/2rPbjSC.
24. Ianonne, op cit., 152.
25. http://bit.ly/2soITgE.
26. http://bit.ly/2rPbjSC.
27. http://bit.ly/2rU6eDL.
28. http://bit.ly/2tMnlbQ.
29. Hans Belting, "An Image and Its Function in the Liturgy: The Man of Sorrows in Byzantium," Dumbarton Oaks Papers, Vol. 34-35 (1980-81) 1-16. Belting, *The Image and Its Public in the Middle Ages* (New Rochelle, NY: Aristide Caratzas, 1981).
30. Ernst Kitzinger, "The Cult of Images in the Age before Iconoclasm," Dumbarton Oaks Papers, Vol. 8, (1954), 113.
31. Glanville Downey, *A History of Antioch in Syria* (Princeton, NJ: Princeton 1961), 554.

32. Mark Guscin, *The Tradition of the Image of Edessa* (Newcastle Upon Tyne: Cambridge Scholars Publishing, 2016), 17.
33. Thierry Lenain, *Art Forgery* (London: Reaktion Books, 2011), 144.
34. Guscin, *The Tradition of the Image of Edessa*, op cit., 2.
35. Ibid., 2.
36. Ibid., 17.
37. Andrea Nicolotti, *From the Mandylion of Edessa to the Shroud of Turin* (Leiden & Boston: Brill, 2014), 27.
38. Ibid., 26.
39. A.M. DUBARLE, *Histoire ancienne du linceul de Turin*, pp. 83-84. http://bit.ly/2rPbjSC.
40. Linda Cooper, "The eleventh century Old French Life of Saint Alexis," *Modern Philology*, August 1986, 1ff.
41. Kitzinger, op cit., 84.
42. Averil Cameron, "The Sceptic and the Shroud," Inaugural Lecture in the Departments of Classics and History delivered at King's College London, 29 April 1980, 8-9.
43. Ian Wilson, "The Shroud and the Mandylion," Lecture given at the Centre of Asian Studies, University of Hong Kong, March 3, 1986.
44. http://bit.ly/2trPwx8.
45. Guscin, *The Image of Edessa*, op cit., 226.
46. Gilbert R. Lavoie, M.D., *Unlocking the Secrets of the Shroud* (Allen, Texas: Thomas More Publishing, 1998), 65-6.
47. *The Image and Its Public in the Middle Ages* (New York: 1990).
48. http://bit.ly/2trIN6J.
49. Mark Guscin, "The Sermon of Gregory Referendarius." http://bit.ly/2trvpzh.
50. "The Palace Revolution of John Comnenus Nicholas Mesarites," http://bit.ly/2szJOJy.
51. Peter F. Dembowski "Sindon in the Old French Chronicle of Robert De Clari. " http://bit.ly/2tsaYT5.
52. De Wesselow, op cit., 175.
53. Ibid., 176.
54. Ibid., 177.
55. Ibid., 177.
56. Ibid., 178 ff.

57 http://bit.ly/2trQERD.
58 Ian Wilson, *The Shroud* (London: Bantam, 2010), 277.
59 http://bit.ly/2rPtwzk.
60 http://bit.ly/2trQER^{59}D.
61 De Wesselow, op cit., 182.
62 Ibid., 183.
63 http://bit.ly/2soXeKh.
64 http://bit.ly/2sUxB4Z.
65 http://bit.ly/2trQERD.
66 http://bit.ly/2rPYHut.
67 http://bit.ly/2szwYLp.
68 De Wesselow, op cit., 188.

2.8 Antiquity and Authenticity
1 http://bbc.in/2rTS9pC).
2 http://bit.ly/2ssR77t.
3 http://bit.ly/2tsg8hU.

III Who Was the Man of the Shroud?

1 http://go.nature.com/2rUnKrk.
2 Schafersman, S.D., 1982, "Science, the public, and the Shroud of Turin," *The Skeptical Inquirer*, Vol. 6, No. 3, Spring,.42.
3 http://www.infidels.org/library/modern/jeff_lowder/jury/chap5.html.
4 Michael Grant, *Jesus An Historian's Review of the Gospels* (New York: Charles Scribner's Sons, 1977), 200.
5 *The Oxford History of the Biblical World* ed. by Michael Coogan (New York: Oxford University Press, 1998), 528.
6 http://bit.ly/2sSWWfa.
7 "What Life Means to Einstein," An Interview by George Sylvester Viereck,*The Saturday Evening Post*, Oct. 26, 1929, 17.
8 James V. Morrison, Review of *What are the Gospels?*, *Bryn Mawr Classical Review*, May 31, 2005, 188, 193.
9 Richard Burridge, *What Are the Gospels? A Comparison with Greco-Roman Biography* (Grand Rapids: Eerdmans, 2004), 304.

[10] Paula Fredericksen, "Introduction – The History of the Historical Jesus," *Jesus of Nazareth King of the Jews: A Jewish Life and the Emergence of Christianity* (Vintage Press, 2000), 8.
[11] Diarmaid MacCulloch, *A History of Christianity*, (London: Penguin, 2010), 92.
[12] "The Prince of Peace and How He Changed the World," *Parade*, 2016, 79.
[13] Interview with RAV.
[14] Sholem Asch, *One Destiny* (New York: G.P. Putnam's Sons, 1945), 5-8.
[15] Pinchas Lapide, *The Resurrection of Jesus – A Jewish Perspective* (Minneapolis: Augsburg Publishing House), 1983, 125-6.
[16] Ibid., 129-130.
[17] Ibid., 130.
[18] https://www.washingtonpost.com/archive/local/2003/04/19/new-book-defends-gospel-account-of-resurrection-story/e7dd69a5-b88c-4351-ad75-fd8e62db2515/.
[19] N.T. Wright, The Faraday Lecture 2007, Cambridge, May 15 2007.
[20] Ludwig Wittgenstein, *Culture and Value*, transl. Peter Winch (Oxford: Blackwell, 1980), 33.
[21] http://bit.ly/2sUxB4Z.
[22] Herbert Thurston, *The Month*, Volume 101, January-June 1903, 19.
[23] John C. Iannone, op cit., 64.
[24] http://bit.ly/2sWQFjC.

IV Shroud in a Shroud

[1] C.D. Broad, *The Mind and Its Place in Nature* (London: Kegan Paul, 1925), 623.
[2] http://bit.ly/2m09ew.
[3] George Orwell, Notes on Nationalism, May, 1945. http://bit.ly/1QiVrwG.
[4] "A Study of the Kirlian Effect," by Arleen Watkins and William Bickel in *The Hundredth Monkey* ed by Kendrick Frazier (Amherst, New York: Prometheus Books, 1991), 221.

5   http://bit.ly/2sTUaHk.
6   De Wesselow, op cit., 139.
7   http://bit.ly/2rP4Bfr.
8   https://bit.ly/2WiqlLV.
9   https://bit.ly/2MvDrB5.
10 https://nbcnews.to/2SOIEH2.
11 http://bbc.in/2szLCCu.
12 http://bit.ly/2sUVIAB.
13 http://go.nature.com/2rUnKrk.
14 http://bit.ly/2sTUaHk.

V Shroud Unshrouded

1   http://bit.ly/2rUl6Se.
2   John C. Iannone, op cit., 16.
3   http://bit.ly/2sTUaHk.
4   http://bit.ly/1DrhOuA.
5   For the use of skin bacteria for identification purposes, see Fierer N, Lauber CL, Zhou N, McDonald D, Costello EK, Knight R, "Forensic identification using skin bacterial communities," Proc Natl Acad Sci USA, 2010 Apr 6; 107 (14): 6477-81.
6   http://bit.ly/2uqKcgO.

Appendix 1
Photographs of the Microbiome Experiments

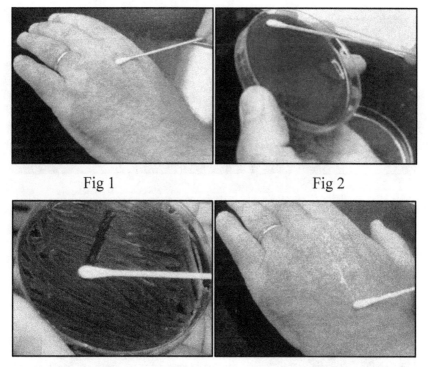

Fig 1

Fig 2

Fig 3

Fig 4

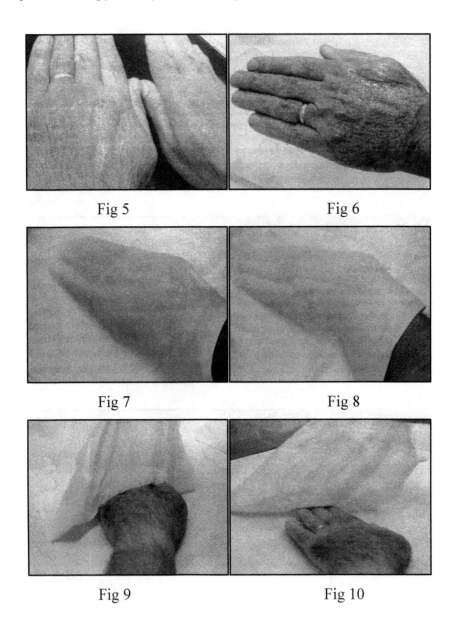

Fig 5

Fig 6

Fig 7

Fig 8

Fig 9

Fig 10

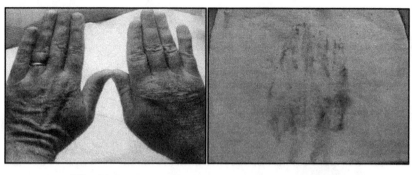

Fig 11　　　　　　　　　　Fig 12

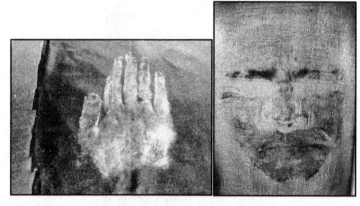

Fig 13　　　　　　　　　　Fig 14

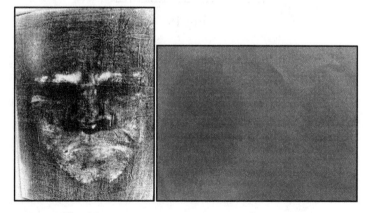

Fig 15　　　　　　　　　　Fig 16

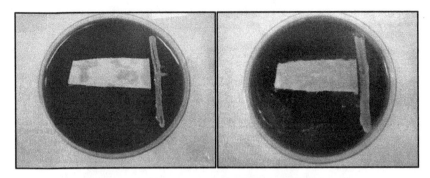

Fig 17   Fig 18

Fig 19

Appendix 2
Historical Documents supporting the Presence of the Shroud in Constantinople (from Professor Daniel C. Scavone)

(reprinted here with the kind permission of Professor Daniel C. Scavone)

DOCUMENT I. THE *NARRATIO DE IMAGINE EDESSENA* 944

On August 15, 944, amidst great celebrations, the Mandylion arrived in Constantinople from Edessa. It was still stretched out against a board and sealed inside its oblong case, the face visible in the circular central opening, as it was subsequently seen by artists who made copies of it. ...

What interests us now is Constantine's personal description of the image: It was extremely faint, more like a "moist secretion without pigment or the painter's art." Equally curious–and increasingly significant in light of Documents III and IV–is a second version of the origin of the Edessa cloth which comes later in this same *Narratio* and which Constantine says he preferred:

"There is another story: ... When Christ was about to go voluntarily to death ... sweat dripped from him like drops of blood. Then they say he took this piece of cloth which we see now from one of the disciples and wiped off the drops of sweat on it." This version would be inexplicable unless we suppose that traces of blood were noticed on the face."

DOCUMENT II. SYMEON MAGISTER 944

The *Narratio*'s account of a nearly imperceptible image is corroborated and embellished by Symeon Magister, writing his *Chronographia* also in the tenth century and likely also under the influence of Constantine VII. He asserts that while Constantine could see the faint image in its details (eyes and ears: ophthαlμoύς και οτα) his two brothers-in-law and rivals for the throne could barely make out an outline.

DOCUMENT III. GREGORY REFERENDARIUS 944

As recently as 1986 a Rome classicist, G. Zaninotto, turned up in the Vatican Archives a 17-page Greek text (*Codex Vaticanus Graecus* 511) of a sermon delivered by one Gregory, Archdeacon and *referendarius* of Hagia Sophia in Constantinople, on August 16, 944, the day after the Mandylion's arrival. As an eyewitness of the events, Gregory again recites the original Abgar legend, and he describes the image as formed by "the perspiration of death on [Jesus'] face." Then comes the most arresting part: he speaks of the wound in Jesus' side (πλευρα) and the blood and water found there (haιμα και hydor eκεi):

"[This image of Christ] was imprinted only by the perspiration of the agony running down the face of the Prince of life as clots of blood drawn by the finger of God.... And the portrait ...has been embellished by the drops from his own side. The two things are full of instruction: blood and water there, and here the perspiration and figure. The realities are equal for they derive from one and the same being.... teaching that the perspiration which formed the image and which made the side to bleed were of the same nature that formed the portrait."

Describing the Edessa cloth, then, Gregory has divulged that it might have contained more than a facial image. ...

DOCUMENT IV. LETTER OF CONSTANTINE VII 958

A letter of the same Constantine VII to encourage his troops campaigning around Tarsus in 958 is the first explicit introduction of the burial shroud icon of Jesus in this context. The letter announced that the Emperor was sending a supply of holy water consecrated by contact with the relics of Christ's Passion which were then in the capital. No mention is made of the recently acquired Mandylion: as a relic of Jesus' ministry it would have been out of place among the relics of the Passion. Reference is made, however, to the precious wood [of the cross], the unstained lance, the precious inscription [probably the *titulus* attached to the cross], the reed which caused miracles, the life-giving blood from his side, the venerable tunic, the sacred linens (σπάργανα), the *sindon* which God wore, and other symbols of the immaculate Passion.[20]

The term used here for "sacred linens," *spargana*, usually means infant's "swaddling cloths," but here must mean burial linens, as it does in several other texts. The precise identity of this *sindon* has been enigmatic, since no mention exists of the arrival in the capital of Jesus' burial sheet, but it acquires some clarity with Zaninotto's recovery of Doc. III. Just as in the Gregory Sermon, the words of this text may suggest that the Byzantines could see "blood" from the side of the figure depicted on a cloth.

Document III is strong evidence that the Edessa icon was indeed a larger object, harmonious with the words *sindon* and *tetradiplon* of the *Acts of Thaddeus*, and was seen to be stained red in the correct places. It must thus have been unfolded in Constantinople sometime after its arrival in 944. A possible unfolding is evidenced by the imperial letter of 958 (Doc. IV), where suddenly, without fanfare, Jesus' *sindon* is first announced. At the time of its arrival in 944, the status of the Edessa icon must, it seems, be understood as follows: Still enframed or encased as described earlier and as seen by artists, and still generally considered to be the towel of the Abgar narratives, and in the treasury of the Byzantine emperors it was inaccessible to the public (as it had been in Edessa). Its size (larger and folded in eight layers) and nature were not fully known and not often pondered. Certainly its possible identity as Jesus' bloody burial wrapping was not immediately recognized or, if it was, then by only a few intimates and not generally broadcast. The Byzantines were too much under the spell of the Abgar cycle to have considered the implications of the side-wound. ...

DOCUMENT VI. ALEXIUS I COMNENUS

A letter which bears the date 1095 falls next under our purview. It purports to be an invitation sent by Byzantine Emperor Alexius I Comnenus (1081-1118) to his friend Robert the Frisian, Count of the Flemings (1071-1093) and to all the princes of the realm. ... [In the letter Alexius states that] the city houses great treasures as well as the precious relics of the Lord. These are then named, and include, unequivocally for the first time in these sources, "the linen cloths found in the sepulchre after his resurrection."

To dismiss this letter as a spurious piece of Latin propaganda virtually making the Byzantine emperor beg for the Latins' expropriation of the imperial relics during the Fourth Crusade is to miss its significance as a Byzantine document referring to the presence of Jesus' burial wrappings in Constantinople. Indeed, were it not for the enigmatic Document IV, this letter would be the first such reference. Most historians have agreed that Alexius would not have written such words, but they also concur that this *epistula* probably "depends on an authentic letter of the basileus" written with another end in mind and that it dates, variously, from 1091 to 1105. ...

Kurt Weitzmann and Hans Belting have shown that by c. 1100 Byzantine iconography had evolved a new style in the depiction of the events of Easter: the *threnos* or "Lamentation." The contemporaneity of this *epistula* and the developed *threnos* art in Byzantium is striking, for thus it signals with a twin corroboration what the large burial cloth icon of Christ must have looked like. Jesus is now shown lying upon a full-length shroud after being removed from the cross; in many examples he is naked and with hands folded upon his abdomen or over his loins. In addition to this new mural art, Byzantine *epitaphioi* or embroidered cloth, symbolizing Jesus' shroud in the Good Friday liturgy, show Jesus in full-length, i.e., in the *threnos* attitude. ...

A group of Latin texts should be considered. ... Somehow–via returning crusaders?–the Abgar story became quite popular in the West in the 12th c. ... [Two] western writers of the Abgar story provide an important clue in the emergent and widening awareness both that the Edessa cloth was larger than originally thought and that it contained a full-body image of Jesus. They are the *Ecclesiastical History* of the English monk Ordericus Vitalis, ca. 1141; and the *otia imperialia* of Gervase of Tilbury, ca. 1211.

In Edessa, it seems, the image was always described as a face only. What is remarkable about these Latin Abgar accounts is the fact that in all of them, what Abgar received was not just a facial image, but one which enabled the viewer to discern the form and stature of Jesus' entire body.[26] If they truly derive from a lost Syriac original from Edessa's archives, as they claim, each one drawing its claim from its own immediate source, they open the possibility (only hinted at in the sources) that already in Edessa someone had known that the Mandylion was an icon of Christ's entire body.

The clue leading to the conclusion that the lost Syriac original used by the western sources was written before the Mandylion left Edessa in 944 is the line in all three Latin texts that "this linen from antiquity still remains uncorrupted in Syrian Mesopotamia in Edessa." ...

Again, the descriptions of the image, no longer as face-only but now as entire body, relate chronologically to (a) the emergent *threnos* and *epitaphios* scenes in the East, which themselves suggest b) an awareness of an imaged shroud of Jesus, and (c) could be witnessed by Western Crusaders in Byzantine churches. Yet only in our Document VI was the much larger and imaged Mandylion recognized as a burial sindon. The Abgar/Mandylion mind-set retained its hold on the authors of the Latin versions, even while they (and possibly their original Syriac source-text) had altered the legend in a significant manner.

Gervase of Tilbury had certainly heard of a cloth bearing the full image of Jesus. Besides giving the old Abgar/Edessa version, he even gives a second account of a bloodied full-body image on a cloth, this time in a context related to the burial of Christ; it has no parallel in Byzantium, to my knowledge; but it is *acheiropoietos*. He writes:

There is another figure of the Lord expressed on cloth which has its origin in *Gestis de Vultu Lucano* (the events surrounding the Volto Santo of Lucca). When the Lord our Redeemer, hung from the cross stripped of his clothing, Joseph of Arimathea approached Mary, the mother of the Lord, and the other women who had followed the Lord in His Passion, and said: Do you love Him so little that you allow him to hang there naked and not do anything about it? Moved by this castigation, the mother and the others with her bought a spotless *linteum* so ample and large that it covered the whole body, and when He was taken down the image of the whole body hanging from the cross appeared expressed on the linen.[27]

Note that it is certain from his writings that Gervase never saw the actual cloth.

Deriving, as they claim, from a pre-944 Edessan text, all three Latin texts include information about the rituals associated with the image when it was in Edessa. And this lends credence to their claim of a Syriac model. Most notably, they state that the cloth with full-body image was kept in a gold chest (*scrinium*) and that: [when displayed] on Easter it used to change its appearance according to different ages, that is, it showed itself in infancy at the first hour of the day, childhood at the third hour, adolescence at the sixth hour, and the fullness of age at the ninth hour, when the Son of God came to His Passion for the weight of our sins and endured the awful sacrifice of the cross.[28]

Oddly this did not appear in the "Liturgical Tractate" where Edessan rituals were earlier described. What can these words mean? The most acceptable answer is one that harmonizes with two other eyewitness descriptions of the cloth in Documents XI and XII. Accepting from the texts already discussed that Edessa's cloth bore a faint painting of an entire body, we may infer from Documents XI and XII that the image of the full and bloodstained body was revealed gradually by the unfolding of the cloth in sections, beginning with the feet and lastly showing the whole bloodstained body. The comparison of the gradually unfolded increments of the body with successive periods of Christ's life would thus have been symbolic, part of the belief-system of the Edessenes.

DOCUMENT VII. ENGLISH PILGRIM 1150.

The Edessa cloth with facial image is not mentioned in Constantinople again until 1150 by an English pilgrim to

Constantinople. He saw what he describes as a gold container, *capsula aurea*, in which "is the *mantile* which, applied to the Lord's face, retained the image of his face."[31] He also mentions the "*sudarium* which was over his head." It is yet another reference to a funerary cloth of Jesus in Constantinople, though it does not seem to be a body shroud.[32] This and the following three documents continue the confusion that thwarts one's efforts to identify the precise objects in the imperial relic collection.

DOCUMENT VIII. NICHOLAS SOEMUNDARSON 1157.

Seven years later (1157) this confusion of terms continues when Nicholas Soemundarson (Thingeyrensis), an Icelandic pilgrim, wrote in his native Icelandic his very detailed inventory of the palace relics. Riant has given us a Latin translation of Nicholas' Icelandic: "*fasciae* with *sudarium* and blood of Christ." Nicholas made no mention of the frame or box holding the cloth of Edessa, and indeed, the reference to blood demands that we interpret these as Passion cloths. Meanwhile, as between *fasciae* ("bands"), as distinguished from *sudarium,* both Latin translations from Icelandic, it is possible but not certain that one of the terms may denote a larger body cloth.[33]

DOCUMENTS IX and X. WILLIAM OF TYRE; ANTONIUS OF NOVGOROD

In 1171 Archbishop William of Tyre was admitted, he says, into the imperial treasury, where saw the *syndon* of Christ. This is the ordinary New Testament word for a body shroud and its sometime use in these contexts to denote the Edessa cloth seems only to hint further that either the Edessa cloth was larger than a face-towel or that another cloth, large and bloodstained, was present in the treasury.[34] After this time, both the Edessa cloth and the burial linens regularly appear in the same inventories.

In 1200 the inventory of Antonius of Novgorod similarly names two linen cloths: *linteum* and "*linteum* representing the face of Christ."[35] Recall that earlier documents had tended towards the conclusion that the Edessa cloth was large (*tetradiplon*) and bloodied, and therefore might be identical with that cloth reputed in the inventories to be the burial wrapping of Jesus. The text of Antonius does nothing to elucidate those conclusions.

DOCUMENT XI. NICHOLAS MESARITES 1201

The plot thickens when Nicholas Mesarites, in 1201 the *skeuophylax* (overseer) of the treasuries in the Pharos Chapel of the Boucoleon Palace of the emperors in Constantinople, again describes two separate objects. One is the Burial *sindones* of Christ: these are of linen. They are of cheap and easy to find material, and defying destruction since they wrapped the uncircumscribed, fragrant-with-myrrh, naked body after the Passion... . In this place He rises again and the sudarium and the burial sindons can prove it ...

The words of this eyewitness intimate that he had seen a naked man's image on one of these cloths. His use of the word *aperileipton*, "uncircumscribed," suggests that this image was lacking an outline. It could also be rendered as "uncontainable," meaning that the limitless spiritual nature of God had somehow been contained in these cloths at the time when Jesus' body was wrapped inside them. His reference to the Passion implies the visible presence of blood on the cloth. Without too great a stretch, Mesarites' words provide us an eyewitness confirmation of the hints developed from so many other documents already discussed.

Nicholas, however, also specifically mentions as a separate second object in his care the towel (*cheiromaktron*) with a "prototypal" (*prototupw*) image of Jesus on it made "as if by some art of drawing not wrought by hand (*acheiropoietw*)."[36]

again So any absolute confirmation of the identification (made possible by the Gregory sermon, Document III, et al) of the Edessan Mandylion (facial image only) and shroud of Jesus (whole body image with presence of visible blood and water from the side wound) remains elusive.

DOCUMENT XII. ROBERT OF CLARI 1203: CLARITY AT LAST

A burial *sydoines* certainly bearing the figure of the Lord is described in the Church of Our Lady of Blachernae by Robert of Clari, knight of the Fourth Crusade on tour in Constantinople in 1203-04. This passage has long been regarded by scholars of the Turin Shroud as the *locus classicus* attesting the presence in the Eastern capital of that famous Shroud:

There was another of the churches which they called My Lady Saint Mary of Blachernae, where was kept the *sydoines* in which Our Lord had been wrapped, which stood up straight every Friday so that the features of Our Lord could be plainly seen there. And no one, either Greek or French, ever knew what became of this *sydoines* after the city was taken.[37]

Clari also saw elsewhere, in the relic treasury of the Pharos Church of the imperial palace (that treasury in Mesarites' care two years prior), the two *tabulae* or cases which supposedly contained the famous Edessa towel (*touaile*) and the imaged tile (*tiule*).[38] Importantly, Clari never said he had seen the contents of these *tabulae*. Taken with the previous document of Mesarites, the words of Clari are refreshingly supportive. The imaged burial wrapping in Blachernae chapel seems to be identifiable with that which Mesarites protected in the Pharos treasury. Mesarites' words "He rises again" seem paralleled by Clari's "stood straight up," and may refer to its being displayed by being gradually pulled up from its case until one could see the naked and blood-stained body of Christ.

DOCUMENT XIII. NICHOLAS MESARITES 1207

In 1207 the same Nicholas Mesarites, ... In this eulogy Mesarites again refers to Constantinople as possessing the burial wrappings of Jesus, and this reference has been used as evidence that the Shroud was still present in the city in 1207.[41] The latter position breaks down when it is noticed that in fact, Mesarites' words in the *Epitaphios* are largely a direct quote from his 1201 report (Doc. XI) and are used by him here only for rhetorical effect.[42]

1. In both places Mesarites lists the relics of Jesus' Passion, including the burial wrappings.

2. Both texts employ the symetrical contrast of Constantinople and Judaea: the Passion occurred there, but the relics are here.

3. Both texts add, identically, "Why should I go on and on? ... (The Lord himself) is here, as if in the original, his impression stamped in the towel and impressed into the easily broken clay (tile) as if in some graphic art not wrought by hand." ...

DOCUMENT XIV. NICHOLAS OF OTRANTO 1207

In the years immediately after the Latin takeover of Constantinople in 1204, a series of discussions took place between Greek clergy and papal envoys ...One of the interpreters at these meetings, a man fluent in both Latin and Greek, was Nicholas of Otranto, abbot of Casole monastery in southern Italy.

His reference to the shroud of Jesus comes in the midst of his discussion in 1207 ... Among the relics of the Passion, which he now enumerated, were a portion of that bread and Jesus' *spargana,* Greek for "linens." This word normally renders infant's swaddling clothes, and the *fascia* of his Nicholas' Latin translation does not help. Since, however, Nicholas was listing relics of the Passion, he must mean burial linens. Here is the crucial passage:

When the city was captured by the French knights, entering as thieves, even in the treasury of the Great Palace where the holy objects were placed, they found among other things the precious wood, the crown of thorns, the sandals of the Savior, the nail (sic), and the spargana/*fascia* which we (later) saw with our own eyes. ...

DOCUMENT XV. THEODORE ANGELUS' LETTER 1205

he territory of Epirus, however, remained a center of Greek power under Michael Angelus as Despot. Michael and his brother, Theodore, were nephews of Isaac II Angelus, one of three Byzantine Emperors who were deposed during the Fourth Crusade. The document in this instance is a letter dated 1 August 1205 from Theodore in the name of Michael to Pope Innocent III. Here are the pertinent passages.

Theodore Angelus wishes long life for Innocent [III], Lord and Pope at old Rome, in the name of Michael, Lord of Epirus and in his own name. In April of last year a crusading army, having falsely set out to liberate the Holy Land, instead laid waste the city of Constantine. During the sack, troops of Venice and France looted even the holy sanctuaries. The Venetians partitioned the treasures of gold, silver, and ivory while the French did the same with the relics of the saints and the most sacred of all, the linen in which our Lord Jesus Christ was wrapped after his death and before the resurrection. We know that the sacred objects are preserved by their predators in Venice, in France, and in other places, the sacred linen in Athens ...

Rome, Kalends of August, 1205.

DOCUMENT XVI. BALDWIN II: GOLDEN BULL 1247

The last Latin Byzantine Emperor, Baldwin II, was sorely in need of funds to maintain his armies. ... [The] Golden Bull of Baldwin II, ceded all these relics, which are enumerated, to the French King in perpetuity, in consideration for still another loan. ...

[It] lists among the relics ceded to Louis "part of the *sudarium* (*pars sudarii*) in which Christ's body was wrapped in the tomb." ...

Recall that although the imperial letter of 958 (Doc. IV) named a burial cloth, it was not until 1095 (Doc. VI) that the documents began to attest more regularly to a recognition of the burial cloth in the capital. Both Mesarites and Clari appear to corroborate what the cumulative documents from the 6th to the 12th c. suggest: that the Edessa cloth was eventually unframed and discovered to hold an impression of the entire and bloodied body of Jesus. That which came to known as the *toella in tabula inserta* would then and logically be a copy of Edessa's *Mandylion* as it had appeared—i.e., the face only of Jesus—upon its arrival in Constantinople.

To sum up the points made in this paper: a linen cloth or cloths described as the burial wrappings of Jesus are attested in many Constantinople documents from 944 to 1203, twice with his image if one counts Mesarites (Doc. XI), and several times described as bloodied. No record exists of the arrival of Jesus' burial cloth in the capital, and no celebration such as accompanied the Edessa cloth in 944. Yet it was there. Judging from copious documents and artistic representations made in Constantinople and elsewhere from 944 to 1150, the Edessa towel always with the image of Jesus' face may be identical with Jesus' Shroud in folded form, enclosed in a case with face exposed. Before that, from at latest 544 to 944, this cloth was certainly in Edessa. If the Edessa cloth and Jesus' purported shroud are indeed one and the same object, that assumed burial cloth may have a pedigree back at least to 544, and if the Abgar legend has any historical worth, to the 4th c. and even, accepting the descriptive evidence, to the very time of Christ. If the pieces of this elaborate puzzle truly fit as they seem to, the blood-stained burial cloth with faint unpainted image would have a documented history back to palaeochristianity and may in fact be the actual tomb wrapping of Jesus.

"Acheiropoietos Jesus images in Constantinople: the Documentary Evidence," http://bit.ly/2rPtwzk.
Daniel C. Scavone, University of Southern Indiana (6/21/96; 4/23/01; 2-25-04; 11-24-04, 12-02-05, 01-03-06, 10-07-2006)